Economics, Ecology, and Policy for the Bioeconomy

This book demonstrates that a holistic approach to the bioeconomy is essential if it is to achieve its full potential in driving economic growth while simultaneously providing ecological, social, and technological benefits.

Definitions of the "bioeconomy" vary, but in general it incorporates the ways in which societies manage and distribute their primary or secondary biological resources for further use in everyday life (e.g., food, materials, and energy). The classical sectors related to the bioeconomy have therefore been agriculture, forestry, and aquaculture, now extended to include bioenergy, biofuels, biochemicals, and other processing and service industries. There are also related new concepts such us the blue economy, the green economy, and the circular economy. This book integrates these definitions, sectoral analyses, and new concepts into a fully rounded study of the bioeconomy. It is argued that the key aims in the coming years have to be the harmonisation of public policies between different sectors, regulation of legislative framework for the bioeconomy, and clear communication of these issues. In particular, the book argues that a strengthening of the monitoring and evaluation of the impacts of the bioeconomy on society are an essential starting point. For this to be effective, appropriate indicators need to be established and defined for the monitoring of the effects of these resilient policies related to the bioeconomy and their impact on local and regional development and quality of life.

This book is essential reading for anyone interested in the bioeconomy including students and scholars of ecological economics, environmental economics, sustainability, innovation, and regional development.

Sanja Tišma, Director of the Institute for Development and International Relations (IRMO), Zagreb, Croatia.

Anamarija Farkaš, researcher at the Institute for Development and International Relations (IRMO), Zagreb, Croatia.

Anamarija Pisarović, researcher at the Institute for Development and International Relations (IRMO), Zagreb, Croatia.

Marina Funduk, researcher at the Institute for Development and International Relations (IRMO), Zagreb, Croatia.

Iva Tolić, researcher at the Institute for Development and International Relations (IRMO), Zagreb, Croatia.

Routledge Studies in Ecological Economics

Physical Limits to Economic Growth
Perspectives of Economic, Social, and Complexity Science
Edited by Roberto Burlando and Angelo Tartaglia

Feminist Political Ecology and the Economics of Care
In Search of Economic Alternatives
Edited by Christine Bauhardt and Wendy Harcourt

Anarchism and Ecological Economics
A Political Platform for Ecological Economics
Ove Daniel Jakobsen

Water Resources and Economic Processes
Edited by Tiziano Distefano

The Degrowth Alternative
A Path to Address our Environmental Crisis?
Diana Stuart, Ryan Gunderson and Brian Petersen

A History of Ecological Economic Thought
Marco P. Vianna Franco and Antoine Missemer

Radical Ecological Economics and Accounting to Save the Planet
The Failure of Mainstream Economists
Jacques Richard

The Informal Sector and the Environment
Edited by Ranjula Bali Swain and Uma Kambhampati

Economics, Ecology, and Policy for the Bioeconomy
A Holistic Approach
Sanja Tišma, Anamarija Farkaš, Anamarija Pisarović, Marina Funduk, and Iva Tolić

For more information about this series, please visit: *www.routledge.com/series/RSEE*

Economics, Ecology, and Policy for the Bioeconomy

A Holistic Approach

Sanja Tišma, Anamarija Farkaš, Anamarija Pisarović, Marina Funduk, and Iva Tolić

Routledge
Taylor & Francis Group
LONDON AND NEW YORK

First published 2023
by Routledge
4 Park Square, Milton Park, Abingdon, Oxon OX14 4RN

and by Routledge
605 Third Avenue, New York, NY 10158

Routledge is an imprint of the Taylor & Francis Group, an informa business

British Library Cataloguing-in-Publication Data
A catalogue record for this book is available from the British Library

Library of Congress Cataloging-in-Publication Data
A catalog record for this book has been requested

ISBN: 978-1-032-12241-0 (hbk)
ISBN: 978-1-032-12242-7 (pbk)
ISBN: 978-1-003-22373-3 (ebk)

DOI: 10.4324/9781003223733

Typeset in Bembo
by Apex CoVantage, LLC

Contents

Figures

Abbreviations

CAGR – compound annual growth rate
EU – European Union
FAO – Food and Agriculture Organization of the United Nations
GDP – gross domestic product
GHG – greenhouse gas
ILUC – indirect land-use change
JRC – Joint Research Centre
LCA – life-cycle assessment
LCSA – life-cycle sustainability assessment
MAGNET – Modular Applied GeNeral Equilibrium Tool
OECD – Organisation for Economic Co-operation and Development
R&D – research and development
RED – Renewable Energy Directive
RED II – revised Renewable Energy Directive 2018/2001/EU
SDG – Sustainable Development Goal
SME – small and medium-sized enterprises
UN – United Nations

Introduction

Sanja Tišma

Although there has been a range of research and strategic and planning policy documents related to the bioeconomy, there is still no unique definition of this important scientific and professional topic, nor is there an interdisciplinary, holistic approach to its study. Generally, the bioeconomy is the economy that analyses how societies manage and distribute their primary or secondary biological resources for further use in everyday life, namely for food, materials, and energy. The bioeconomy has also been defined as "the production, utilisation and conservation of biological resources, including related knowledge, science, technology, and innovation, to provide information, products, processes, and services in all economic sectors aiming toward a sustainable economy" (International Advisory Council of the Global Bioeconomy Summit 2018 2018, p. 4). Besides the classical sectors related to the bioeconomy such as agriculture, forestry, and aquaculture, the approach has been extended to include modern sectors such as bioenergy, biofuels, biochemicals, along with other processing and service industries (i.e., food, paper, textiles, building and construction, chemistry, and biopharmacy). In addition, there are new concepts such as blue economy, green economy, innovation and research, and circular economy.

These sectors and concepts have developed the most in recent years and have been promoted by different sustainability strategies.

The importance of the policy context of the bioeconomy is visible in the example of the European Union (EU). The highest productivity in the bioeconomy in the EU is realised in the generation of bioenergy from biomass but is mainly due to state aids for priority purchase of electricity from biomass. In addition, liquid biofuel and agriculture sectors, and partially aquaculture activities, receive significant state aid. However, such an approach is being abandoned to achieve equal rights in biomass demand for all sectors of the bioeconomy. This example shows that the critical challenge of the development of a bioeconomy today is a classically linear, mono-sectoral approach, which disables a holistic perception of a bioeconomy as a very integrated and cross-sectoral concept. Such a holistic approach to the bioeconomy is a key theme of this book.

Besides analysing the theoretical concept of bioeconomy, related disciplines, and related policy instruments, this book provides a framework of the classical

DOI: 10.4324/9781003223733-1

sector approach to the bioeconomy. In addition, new approaches and sectors based on the bioeconomy are described.

Political factors strongly influence the bioeconomy through the lack of coordination among public bodies included in the bioeconomy activities, lack of communication between public and private sectors, slow introduction of changes, and low awareness of this topic. Therefore, the key aims for the next period refer to the harmonisation of the sectoral public policies, regulation of legislative framework for the field of bioeconomy, and clear and intensive communication of these topics with a broader and interested public. This book should contribute to coping with these challenges.

Chapter 1 provides a theoretical review of the current definitions and concepts of the bioeconomy, mainly used in international documents. Although the term "bioeconomy" has become mainstream in policy papers and strategies globally, currently, there is no common definition for the political concept of bioeconomy. Also, there is no holistic approach in the study of this concept. That is the reason for the somewhat repetitive nature of various definitions of the term in other book chapters. However, it should be stressed that the bioeconomy is not a static notion, and its meaning is continually evolving. On the global level, the bioeconomy is defined as the economy that analyses how societies manage and distribute their primary or secondary biological resources for further use in everyday life, namely for food, materials, and energy. On the level of the EU, the bioeconomy is defined in a somewhat narrower sense as an innovative, low-emissions economy, which ensures the sustainability of agriculture and fisheries, food security, and the sustainable use of renewable biological resources (biomass) for industrial purposes while protecting biodiversity and the environment (European Commission 2012). Accordingly, the EU Bioeconomy Strategy has five goals: ensuring food supply, sustainably managing natural resources, reducing dependence on unrenewable resources, and mitigating and adjusting to climatic changes, as well as creating new jobs and maintaining European competitiveness. Additionally, in their national strategies, all EU member states can determine their priorities depending on the specificities of each country.

The bioeconomy in the EU is perceived as a framework for economic, social, ecological, and technological breakthroughs to implement the European Green Deal. The starting point is finding new ways of producing and consuming resources while respecting the available resources, particularly transitioning from linear economic development based on the exhaustion of fossil and mineral resources towards achieving the UN Sustainable Development Goals (SDGs) and a circular economy.

Chapter 1 considers the complexity of a range of issues and new development concepts used to provide readers with the guidelines for sustainability. Besides providing fundamental bioeconomy theories and concepts, it brings a detailed analysis of related concepts (i.e., sustainable development, circular economy, circular bioeconomy, linear bioeconomy, green and blue economy), comparing them to the concept of bioeconomy. It also emphasises the importance of innovations for breakthroughs in the related fields, stresses key interrelationships and overlapping concepts, and explains common starting points and breakthroughs.

Chapter 2 analyses debates, strategic documents, and action plans related to developing concepts and policies concerning the bioeconomy. It also shows the historical development of these reflections. In addition, it analyses global processes and accompanying documents, such as the results and achievements of the 2020 Bioeconomy Global Forum. This chapter shines a light on critical changes expected to bring about the intensive development of the bioeconomy in the EU through the implementation of the Green Deal. By realising this plan that emphasises the importance of the bioeconomy, Europe should become climate-neutral by the year 2050. A 2018 revision of the 2012 EU Bioeconomy Strategy tried to accelerate the development of the bioeconomy in the EU and link the goals of the bioeconomy to the corresponding UN SDGs. The 2018 update of the Bioeconomy Strategy stresses the importance of a sustainable and circular bioeconomy in achieving strategic goals. It proposes an action plan for the bioeconomy by strengthening the sectors based on biomass, opening investments, and markets, accelerating the development of national bioeconomies throughout Europe, and considering ecological restrictions of the bioeconomy. Many countries have adopted their national strategies of bioeconomy, and there are examples of the countries that have incorporated the bioeconomy in some other sectoral strategies. Thus, the book provides examples of good practices of the countries that have developed their bioeconomy strategic frameworks.

Chapter 3 deals with the bioeconomy from the sectoral perspective. This chapter deals with traditional sectors such as agriculture, forestry, and aquaculture by describing their current state and the new possibilities for development towards more bio-based, "green" practices. Agriculture is essential both for living and for food security. In the bioeconomy, agriculture is perceived through a prism of biological resources, land and crops, plant waste, animal waste, food, materials, and energy chains. Increasing the participation of primary agricultural sectors in the bio-based sector ensures sustainable rural development. The forest-based bioeconomy is focused on the use of wood and non-wood forest products and the forest services such as recreation, tourism, health, and clean environment. The aquaculture-based bioeconomy relies on the blue economy; therefore, it is sometimes called the blue bioeconomy. It includes novel foods and food additives, animal food, nutraceuticals, pharmaceuticals, cosmetics, materials, and energy.

Chapter 4 deals with new, modern sectors that have a more significant impact on the bioeconomy (i.e., bioenergy, biochemistry, and biopharmacy). Through the analysis of particular sectors, concrete examples are provided to point out the possibilities of the practical application of these examples in everyday life. Some examples include challenges of the biorefinery sector, plant production in the agricultural sector, biotechnology and bio-based forestry case studies, and biopharmacy potential.

Chapter 5 analyses the bioeconomy concept through a holistic approach, a novelty aspect this book offers. Key economic factors affecting the development of the bioeconomy are similar to those affecting industrial development in general: lack of knowledgeable and skilful labour, migrations, low productivity, weak financial power, unemployment, low level of availability of funding for investments, expensive loans and/or lack of accessibility of the credits for cofounding of projects, restrictive tax policy, low-level technological development, and excessively long periods of investment return and revenue generation. Ecological factors affecting the development of the bioeconomy are climate changes, pollution, and exploitation of natural resources. These factors have additionally increased during the Covid-19 crisis. Critical social factors are low educational level, mainly among the rural population, low level of ecological awareness, low motivation for work, general pessimism, and low living standards. Technological factors related to the bioeconomy are the high price of new technologies, insufficient patents, and insufficient financial sources for technological development. It should be stressed that the EU Joint Research Centre (JRC), Sevilla, Spain, and the Food and Agriculture Organization (FAO) of the UN are currently working on a new methodology and indicators for monitoring the development of the bioeconomy which will be a starting point for statistical monitoring in the future. However, it is a fact that the bioeconomy is hampered by a lack of statistics on emerging partially bio-based sectors.

The socio-economic indicators monitoring the bioeconomy have been researched to some extent. However, most research is focused on the contribution of the bioeconomy sectors to gross domestic product (GDP) and employment. At the same time, there is a lack of addressing social, environmental, and particularly technological effects of the bioeconomy. Technological results are very important for future innovations and scientific research support investments. Accordingly, due to the accelerated digitalisation in all spheres of human life and work, these impacts of the bioeconomy development should be elaborated. The research conducted so far indicates that the lack of holistic assessment of the bioeconomy stems from the lack of a homogenous definition, which prevents measuring the relevance of the bioeconomy in different economies. Prioritising the goals of some bioeconomy development strategies supporting different sectors and certain comparative advantages some countries have is of particular significance. Examples are abundant natural resources, investments in innovations, research and development, industrial development, and so on.

This chapter sets out to contribute to the holistic perception of the bioeconomy by describing its potential effects on the well-being of the society while also proposing some of the possible breakthroughs for monitoring and assessing the future considerations of this complex new development paradigm.

This book's conclusion summarises the findings and provides recommendations to global, regional, and local decision-makers. The readers can find essential information about the bioeconomy and a fundamental framework for new business ideas. In light of climate challenges and the current Covid-19 pandemic affecting the world, the bioeconomy is more relevant than ever. The development of the bioeconomy envisages entirely new paradigms in business processes. By-products in manufacturing become new raw materials in optimised production processes through new knowledge and novel green technologies. The ways of green mobility are also changing and improving. All of that strengthens the efficiency of the entire system and its resilience to crisis.

The potential that the bioeconomy has for economic growth and ecological, social, and technological aspects of well-being points out that additional measures should be taken for an in-depth analysis of these topics. One of the more important breakthroughs would be strengthening the monitoring and evaluation of the impacts of the bioeconomy on society. Accordingly, in this case, the starting point should be defining the appropriate indicators to holistically monitor the effects of these resilient policies related to the bioeconomy and their impact on local and regional development and the quality of life of people.

Finally, the bioeconomy can serve as a catalyst for a profound systemic change by searching for new ways to produce and consume resources while simultaneously respecting our planetary boundaries. Thus, a holistic approach to the perception of challenges and potentials of the bioeconomy is an innovation in everyday practice.

Bibliography

European Commission, 2012. Innovating for sustainable growth: A bioeconomy for Europe. Available from: https://eur-lex.europa.eu/legal-content/EN/TXT/?uri=COM:2012:0060:FIN#document1

International Advisory Council of the Global Bioeconomy Summit 2018, 2018. Communiqué: Innovation in the global bioeconomy for sustainable and inclusive transformation and wellbeing, 4. Available from: http://agricultura.gencat.cat/web/.content/de_departament/de02_estadistiques_observatoris/27_butlletins/02_butlletins_nd/documents_nd/fitxers_estatics_nd/2018/0208_2018_RDi_Bioeconomia_Bio-mon-cimera-Berlin-Comunicacio-abril-2018.pdf [Accessed 4 January 2022].

1 Concepts of bioeconomy

Iva Tolić

What is the bioeconomy?

The bioeconomy is by no means a new concept; for thousands of years, humankind has been meeting its needs in terms of food, materials, goods, and energy through the use of raw materials and renewable sources. The "bioeconomy", as we refer to it today, represented the modus operandi until the Industrial Revolution. Then, the invention of the steam engine initiated the transformation of the global economy as the primary sources of energy became non-renewable, which is still the case today. However, awareness of the unsustainable pace of consumption of non-renewable energy sources has led to the demand for their alternatives (Pietzsch and Schurr 2017).

One such alternative is the bioeconomy, an innovative economy with low CO_2 emissions, which ensures the sustainability of agriculture and fisheries, food security and the sustainable industrial use of renewable biological resources (biomass) while protecting biodiversity and the environment (European Commission 2018). The bioeconomy encompasses those areas of the economy that use renewable biological resources from land and sea, such as crops, forests, fish, animals, and microorganisms, to produce food, materials, and energy. Unfortunately, there is currently no consensus on the definition of the bioeconomy. Still, many definitions share similar elements, such as low CO_2 emissions and renewable energy sources (National Academies of Sciences, Engineering, and Medicine 2020). In short, the bioeconomy is based on the use of research and innovation in the biological sciences to create economic activity and public benefit (Birner 2018).

Bugge *et al.* (2016), who analysed scientific papers on the bioeconomy, identified three visions of the bioeconomy: the biotechnology, bioresource, and bioecology visions. The biotechnology vision emphasises the importance of biotechnology research and the application and commercialisation of biotechnology. The bioresource vision focuses on researching and developing biological raw materials in sectors such as agriculture, maritime affairs, forestry, and bioenergy. The bioecology vision emphasises the importance of ecological processes that optimise energy and nutrients, promote biodiversity, and avoid monocultures and soil degradation. Simply put, the biotechnology vision represents

DOI: 10.4324/9781003223733-2

the starting point in the applicability of the bioeconomy as a science, the bioresource vision emphasises the potentials of biological raw materials, while the bioecology vision emphasises the potential for integrated processes and systems. The identified visions of the bioeconomy have different aims; the aims of the biotechnology vision relate to economic growth and job creation, those of the bioresource vision relate to economic growth and sustainability, while the aims of the bioecology vision of the bioeconomy relate to sustainability, biodiversity, ecosystem conservation, and prevention of soil degradation. Although the aims of the bioeconomy differ according to visions, it can generally be concluded that the bioeconomy as science aims to effectively respond to challenges such as climate change, food security, health, industrial restructuring, and energy security. However, as with the definition of the bioeconomy, there is still no consensus on the aims of the bioeconomy, which vary from country to country, depending on resources, specialisation, and direction of economic development (International Advisory Council on Global Bioeconomy 2020).

Emergence of the bioeconomy

Although the bioeconomy has always existed as a concept, the term "bioeconomy" or "bioeconomic science" was first used at the International Congress for Aquaculture and Fish Culture in 1931 in Paris by a member of the Romanian Academy, scientist Grigore Antipa (Bogdana *et al.* 2014). However, Nicholas Georgescu-Roegen is considered to be the father of the bioeconomy. In the 1970s, he developed a novel approach, called bioeconomy, based on the analysis of the economic system as an integral part of the environment and therefore subject to the principles of physics and biology. Georgescu-Roegen defined a bioeconomy as

> a new economy, whose purpose is to conserve resources and to obtain rational control over the development and use of technologies in order for it to serve the true human wants – and not rising profits, warfare, or national prestige. We need an economy of survival or, rather, of hope – the theory or the comprehension of a worldwide economy that is predicated on justice and that allows for the wealth of the earth to be shared equally among its inhabitants now and in the future.
>
> (Vogelpohl and Töller 2021, p. 1)

What the concept of the bioeconomy refers to today – the economic activities related to the development, production, and use of biological products and processes (OECD 2009) (i.e., sectors and systems that rely on biological resources), their functions and principles (European Commission 2018) – is rooted in the work of Enriquez (1998), who describes it as "the creation of a new economic sector, the life sciences". Although the paper does not contain the term "bioeconomy", it represents one of the roots of the concept of the bioeconomy; the advances in biological sciences and biotechnology, which have the potential to transform many industrial production processes. This concept was adopted by the European Union (EU) in the early 2000s, especially by Christian Patermann, the father of the bioeconomy in the EU. He recognised the potentials of the concept of the bioeconomy in response to EU problems and needs of the EU, such as increasing agricultural productivity, particularly for uses other than for food. At the same time, the US started encouraging the concept of the bioeconomy, and many other countries around the world followed suit by developing strategic documents related to the bioeconomy (Birner 2018).

The historical account of the bioeconomy shows that, although the bioeconomy has existed since before the Industrial Revolution, the introduction of the bioeconomy into political and strategic frameworks at the national and international levels is a relatively new concept subject to change and constant improvement. As aforementioned, the definition of the bioeconomy is still not generally accepted, and its components (i.e., the elements that the bioeconomy encompasses) differ from country to country. However, the emergence of the bioeconomy (i.e., recognising the importance of the concept of the bioeconomy at the global level) is a continuation of the global development paradigm known as "sustainable development", which is defined in the UN report "Our Common Future" as the

> development that meets the needs of the present without compromising the ability of future generations to meet their own needs. . .[;] sustainable development is a process of change in which the exploitation of resources, the direction of investments, the orientation of technological development; and institutional change are all in harmony and enhance both current and future potential to meet human needs and aspirations.
>
> (United Nations 1987, pp. 41–43)

It is important to stress that the bioeconomy does not represent a change of direction of the global paradigm from sustainable development. Still, the principles of bioeconomy contribute to sustainable development.

Importance of bioeconomy for social development

The bioeconomy is a concept that is extremely important for the further development of society in several ways. First, the bioeconomy has a substantial impact on global development and environmental challenges (FAO 2016), with the most affected sectors being agriculture, health, and industry (OECD 2009). Directly or indirectly, the principles of bioeconomy significantly affect other sectors and areas of social development.

In agriculture, the application of bioeconomy improves plant and animal diversity, provides access to technologies, and expands the number of firms and research institutes that use biotechnology (OECD 2009). Promoting the use of the bioeconomy in agriculture leads to increased productivity and food availability, thereby reducing hunger and poverty and improving the health of the poor. The bioeconomy is also applied in "urban agriculture" using waste materials used for energy purposes. The development of urban agriculture, which uses waste materials and waste heat for food production, enables food safety because it facilitates producing food throughout the year, regardless of weather conditions (BMEL 2014).

The contribution of the bioeconomy to the health sector stems primarily from the development of new technologies such as regenerative, personalised, and preventive medicine. The application of the bioeconomy in the field of health includes therapeutics, diagnostics, pharmacology, functional foods, nutraceuticals, and medical devices (OECD 2009). Furthermore, the bioeconomy has an indirect effect on health; in addition to reducing exhaust gases and thus air pollution, biofuels, filtration technologies, and improvements in boiler technology protect the public from diseases caused by fine dust.

The application of the bioeconomy in the industrial sector refers primarily to the application of environmentally sustainable technologies such as boosting research on high energy density biofuels. Industrial sectors relevant to the bioeconomy include the chemical industry, the paper and pulp industry, the agriculture and forestry industry, the energy sector, and mechanical and process engineering (OECD 2009). Knowledge spillover through industry sectors can increase the economic and social importance of the bioeconomy, which requires the expertise of the authorities and cooperation between the many administrative authorities responsible for agriculture, education, environment, health, industry, natural resources, and research. Industrial biotechnology based on the use of bio-based raw materials and the application of biotechnological processes provides a significant incentive to move towards an economy based on renewable energy sources. In addition, industrial biotechnology creates the basis for new products and innovative methods for producing biofuels and bio-based products, especially for the chemical, food, paper, and textile sectors (BMEL 2014).

The application of the bioeconomy relating to the production of biofuels also contributes to the creation of sustainable energy. Biofuels are mainly produced from cereals, sugar cane, vegetable oils, and rapeseed oils, representing sustainable energy sources. Biofuel production has grown in the last 20 years, and biofuels are becoming increasingly available and accessible to the public. The concept of biorefineries has been developed to produce bioenergy, biofuels, and bio-based products using renewable resources. Biorefinery operational principles are based on producing bioenergy with products that provide the highest added value and use of efficient resources. Biorefinery operational principles are based on single or multiple use of materials, while the by-products and waste generated during production processes are used to power other production processes (i.e., generate energy). In this way, biorefineries not only contribute to the production of more sustainable energy, and the mitigation of greenhouse gases (GHGs), thus reducing the impact on climate change as well as supporting the principles of "zero waste" (European Commission 2012).

In addition to the production of biofuels, the contribution to energy efficiency in existing technological processes is generated by the production of biofuel components based on thermochemical and biological routes of synthesis, the conversion of ingredients of algae, the use of biogas, and the use of biomethane as a transport fuel (BMEL 2014).

Biofuel production has a broad impact on social development; in addition to reducing environmental pollution and generating sustainable energy, biofuel production can reduce poverty in developing countries by producing raw materials for biofuels, which could help small farmers significantly increase yields and incomes, thereby ensuring long-term poverty reduction. Furthermore, large-scale biofuels cultivation could also provide benefits in employment, skills development, and development of secondary industries (Cotula *et al.* 2008). As aforementioned, applying the bioeconomy concept in agriculture is a way of reducing poverty. Furthermore, the use of new technologies and innovations in agriculture leads to an increase in bioenergy and biofuels production and greater availability, safety, and quality of food due to improved knowledge and increased diversity of food products and channels. Finally, poverty alleviation results from the agricultural efficiency increase, the increases in food availability, and the rise in the share of local food supply (European Commission 2012).

The bioeconomy also affects societal development by applying biotechnology (i.e., technological solutions) to improve water quality (OECD 2009). Biotechnology can contribute to developing more efficient production processes that convert carbon sources more efficiently and have a lower water intensity. It also has an indirect impact on the state of the environment (i.e., reducing pollution by maintaining the purity of water) (European Commission 2012). However, it is essential to emphasise that while biofuel production minimises water pollution, it also consumes large amounts of water, which is important to consider when developing strategic documents related to the bioeconomy (Gheewala *et al.* 2011). In addition to water protection, the bioeconomy can contribute to the general protection of terrestrial ecosystems, forests, halting soil degradation and biodiversity degradation.

Biofuel production could significantly impact the fight against climate change. However, to reduce air pollution caused by CO_2 emissions, it is essential to take the habitat of plants used in biofuel production into account as the effect of fuel production by cleaning carbon-rich habitats leads to increased CO_2 emissions compared to fossil fuel emissions. In contrast, biofuels made from biomass grown on degraded agricultural lands or waste biomass reduce emissions (i.e., the carbon footprint) (Fargione *et al.* 2008). Furthermore, the application of biotechnology results in sustainable production and exploitation of biological resources, which will enable more efficient production and the use of waste and waste substances and materials for production purposes, thereby limiting negative environmental impacts and reducing the dependency on fossil resources. All of this will help mitigate climate change (European Commission 2012).

The bioeconomy potential is evident in conserving seas and marine resources. Marine biotechnology includes aquatic living resources in the broadest sense: fisheries, aquaculture, and blue biotechnology. New, innovative products and services could be offered to markets by promoting further research into marine biodiversity and strengthening marine biotechnology. Components derived from marine organisms through biotechnology are already used in the food, pharmaceutical, cosmetic, and chemical industries. The unexploited potential of the sea is even more considerable because more than 90% of marine biodiversity remains unexplored, offering immense potential for the discovery of new species and applications derived from biotechnologies (European Commission 2012).

The bioeconomy could substantially impact economic growth and development, primarily in the labour market. Above all, this refers to the education system that needs to be adapted to labour market needs, creating new, hybrid university curricula targeted at developing interdisciplinary competencies in a broad range of fields (European Commission 2012). In the future, the bioeconomy and its components will represent an essential share of the labour market in need of both high- and less-educated workers and will encourage entrepreneurship and innovation. Additionally, for countries with favourable endowments of land, labour, and trade conditions, biofuels provide an opportunity to develop new export markets and improve the trade balance (Cotula *et al.* 2008).

Bio-based products (i.e., products derived from materials of biological origin) represent additional benefits of the bioeconomy. The advantages of these products over conventional products range from more sustainable production processes to improved functionalities and features. Examples of such products are enzyme-based detergents that work more efficiently at lower temperatures and save energy and products with improved biodegradability and/or lower toxicity (European Commission 2012).

As a rule, the development of the bioeconomy contributes to food security, climate change mitigation, and the management of natural resources in a sustainable way. A sustainable bioeconomy has the potential to address burning environmental and socio-economic issues such as maintaining biodiversity and GHGs balance, reducing dependence on fossil resources, improving social well-being, and creating new jobs. The bioeconomy is an open, innovative, and multidisciplinary approach characterised by cooperation and dialogue between different stakeholders at the global level.

An analysis of the 2018 EU Bioeconomy Strategy showed that the bioeconomy positively impacts various societal goals, including the UN Sustainable Development Goals (SDGs). The goals of the EU Bioeconomy Strategy are linked to 12 of the 17 UN SDGs (Viaggi 2021). However, it is important to stress that the bioeconomy does not in itself contribute to the development of society or the achievement of SDGs. Realising the full potential of the bioeconomy requires regulations, policies, and investments that ensure sustainability; only a sustainable bioeconomy has the potential to help achieve SDGs and support the development of society in general (Hiemann 2019).

Bioeconomy in relation to other sciences

As aforementioned, the bioeconomy is an innovative economy with low CO_2 emissions, which ensures the sustainability of agriculture and fisheries, food security and the sustainable industrial use of renewable biological resources (biomass) while protecting biodiversity and the environment (European Commission 2018). However, the definition of bioeconomy contains elements of other sciences, such as economics, circular economy, green economy, and the like. This raises questions on the differences between the bioeconomy and other sciences (i.e., the position of the bioeconomy in relation to other sciences).

We use economics as a starting point. Economics is a social science that studies how people interact with value, especially the production, distribution, and consumption of goods and services (Krugman and Wells 2012). Simply put, economics is the science that answers the questions of what, how, and for whom something is produced. Economics also observes the behaviour and interactions of economic operators and how economies operate. Conversely, the bioeconomy also answers the questions about what, how, and for whom something is produced but observes a much narrower range of products; products obtained by exploiting renewable natural resources, using the knowledge-based production, biological processes, and laws in an environmentally sound manner. Thus, the **bioeconomy** is a part of **economics**: both sciences observe the production, distribution, and consumption of goods and services; the economy as a whole; and economic operators' behaviour and interaction. The difference between the bioeconomy and economics is that the objectives of the bioeconomy include low levels of CO_2 emissions, sustainable production and consumption, and the protection of biodiversity and the environment. In contrast, the goals of economics refer to economic growth, security, stability, full employment, efficiency, and equality.

The circular economy is a regenerative system in which the input of resources and waste, emissions, and energy loss is minimised by slowing down, closing, and narrowing materials and energy loops. The circular economy can be achieved through long-lasting design, maintenance, repair, reuse, remanufacturing, refurbishing, and recycling (Geissdoerfer *et al.* 2017). Simply put, the core of the circular economy is the production process in which the disposal of all waste materials and resources is the last option. Thus, products are either maintained or collected and reused and remanufactured or recycled. The circular economy also includes fossil resources, minerals, metals, biomass, forest and marine waste, and CO_2 from industrial processes or the atmosphere.

The circular economy strengthens the environmental efficiency of the process and the use of recycled carbon to reduce the use of additional fossil carbon. This is precisely the difference between the circular economy and the bioeconomy, as the bioeconomy replaces fossil carbon with carbon based on biomass from agriculture, forestry, and the marine environment. In addition, the characteristics of the bioeconomy include new advances in agriculture and forestry; new, less toxic processing methods characterised by the use of less invasive chemicals; biotechnology, chemicals, and materials with new properties and functions; and products that are healthier and more natural. The concepts of bioeconomy and circular economy share certain commonalities, such as exploiting improved resources with greater environmental efficiencies and a lower share of GHG emissions, reducing the demand for fossil carbon, and encouraging valorisation of waste and by-products. However, although bioeconomy and circular economy concepts have similar goals and are partially overlapping, neither is entirely part of the other (Carus 2017).

The green economy is a low-carbon, resource-efficient, and socially inclusive economy in which employment and income growth are driven by public and private investment in economic activities, infrastructure, and assets that reduce carbon emissions and pollution, enhance energy and resource efficiency, and prevent the loss of biodiversity and ecosystem services (UNEP 2011). The definition of the concept of the green economy indicates that it includes elements of both the bioeconomy and the circular economy, such as environmental efficiency and the use of renewable energy sources. However, unlike the resource-focused bioeconomy, the green economy acknowledges the underpinning role of all ecological processes (D'Amato *et al.* 2017).

In conclusion, the bioeconomy, the circular economy, and the green economy share the same goal of making development sustainable. However, in reality, these three concepts differ in the level of comprehensiveness. The bioeconomy represents the narrowest concept based on bioproducts and bioenergy. At the same time, the green economy generally refers to renewable energy sources, organic production, recycling processes, and the social acceptability of economic activities. The circular economy is the broadest of the three concepts and deals with the environmental aspect (Beg 2018).

The bioeconomy is an integral part of development plans and development policies, especially in recent years since the importance and contribution of the bioeconomy to the development of society have been recognised. All major world powers, including the EU and the United States, have strategic documents to develop the bioeconomy. Similarly, more countries are developing their strategic plans that would include the introduction of the bioeconomy. Furthermore, the importance of the bioeconomy has been recognised at lower levels, with local units mentioning the bioeconomy as one of the priorities and goals in their development plans and establishing measures for the introduction of the bioeconomy.

Linear and circular bioeconomy

The concepts of bioeconomy and circular economy are similar in that they are based on maintaining the value of products, materials, and resources in the economy for as long as possible. While this may lead to the conclusion that the bioeconomy itself is circular, it is not necessarily the case. The basic bioeconomy processes are linear; they stem from the agriculture and forestry sectors (i.e., bio-innovation in agriculture and forestry and smart agriculture). Because it is based on reproducibility, saving fossil resources, reducing the impact of climate change, and increasing productivity and sustainability, the bioeconomy relies on agriculture and forestry, in addition to traditional food and production needs, in connection with biomass production, as well as waste materials and substances from these two sectors. The following step is the processing of biomass, which is applied in several ways. First, a part of biomass is used to produce chemicals and materials with new properties and functions that are further used to produce bio-based products with new functions and properties. Processed biomass is also used for food and feed production, ensuring food availability and safety. Finally, processed biomass is used as an energy source and a resource for biofuel development, thus achieving greater energy efficiency.

Conversely, the actual goal to be achieved is a circular bioeconomy. Just like the bioeconomy, the concept of a circular bioeconomy has many definitions and various interpretations. Thus, a circular bioeconomy can be interpreted as a framework for reducing dependence on natural resources, transforming manufacturing, promoting sustainable production, using renewable sources, and promoting their conversion into various biologically based products and bioenergy, while growing new jobs and industries (European Commission 2013). A circular bioeconomy is also defined as an idea for stimulating economic growth that combines aspects of circular economy and bioeconomy (Giampietro 2019). The simplest and most applicable definition states that a circular bioeconomy implies more efficient management of bio-based renewable resources by integrating circular economy principles into the bioeconomy (D'Amato *et al.* 2020).

The goal of the circular bioeconomy is to separate economic growth from environmental pressures and dependence on fossil fuels. In fact, the circular bioeconomy contributes to the SDGs, including food security, energy security, renewable energy, clean water, sustainable waste management, environmental protection, climate change reduction, and others (Nagarajan *et al.* 2021).

The concept of circular bioeconomy is based on biorefineries that use biomass, waste biomass, algae, wastewater, organic waste, and CO_2. These resources are subjected to microbial, biochemical, and thermochemical processes to produce various products, such as biofuels, bioplastics, and the like. The processes of converting biomass and other resources into products in biorefineries are characterised by reduced energy consumption, reduced use of raw materials, reduced waste generation, and reduced GHG emissions.

One of the features of the circular bioeconomy entails reusing the products from biorefineries as much as possible. Once the product can no longer be used, it is recycled or remanufactured. If the product is recycled, it becomes a resource used for the operation of biorefineries, which represents the first stage of the circularity of the circular bioeconomy. But if a product is remanufactured, it produces a new product at the beginning of its life cycle, which again supports the circular concept of a circular bioeconomy. Suppose a product cannot be reused, recycled, or remanufactured at the end of its life cycle. In that case, the last step in the circular bioeconomy process refers to the recovery of resources, which are then reused as production inputs in biorefineries. In the concept of a circular bioeconomy, disposal represents the last resort; should it come to it, it is necessary to ensure that disposal is safe for the environment and human health.

Bioeconomy sectors

The bioeconomy is still a relatively new science, and no consensus has been reached on its definition or the sectors it covers. The bioeconomy is in itself not an economic sector but encompasses several sectors, some of which are considered wholly within the bioeconomy (e.g., biotechnological research and development), while some sectors, the so-called mixed, partially included, or hybrid sectors, such as the production of soy printer ink (part of the printing ink manufacturing industry), are only considered as a partial bioeconomy (National Academies of Sciences, Engineering, and Medicine 2020).

The bioeconomy includes the sectors of agriculture; forestry; fisheries; manufacture of food, beverage, and tobacco; manufacture of bio-based textiles; manufacture of wood products and furniture; manufacture of pulp and paper; and chemical, biotechnology, and energy industries (Beluhova-Uzunova *et al.* 2019).

The concept of bioeconomy arose from the idea of replacing raw materials and materials based on fossil resources with biomass, which could be converted into materials, chemicals, and fuels. The bioeconomy was rooted in developing industrial biotechnologies that promised clean energy sources and green industrial processes characterised by less pollution, fewer CO_2 emissions, and less waste. Today, the bioeconomy reconciles economic growth and environmental goals, and the concept has evolved beyond biotechnology (Purnhagen and Matthews 2019). Nevertheless, biotechnology is a sector considered to be wholly within the bioeconomy. But what is biotechnology? The definition of biotechnology is the application of scientific and engineering principles in the processing of materials by biological agents (OECD 2013). Simply put, biotechnology uses living organisms, for instance, intact organisms such as bacteria or natural substances from organisms such as enzymes, to make valuable products. In short, biotechnology uses biological systems and processes to produce useful products and provide services. Today, biotechnology represents a small part of the bioeconomy, mainly as a technology provider. In short, the bioeconomy could be described as "a world where biotechnology contributes to a significant share of economic output" (OECD 2009).

Agriculture is a sector that is essential for life and food security, but it also forms the basis for local entrepreneurship, employment, and social development. The link between agriculture and the bioeconomy is based on the use of renewable resources to produce food, materials, and energy, making resource efficiency and the transition to a low-carbon economy possible (European Investment Bank 2018). In addition to improving plant and animal diversity and increasing food productivity and availability, the application of bioeconomy in agriculture leads to the development of plant varieties from which the main food, feed, and industrial raw materials originate. In addition to genetic modification, biotechnologies include intragenics, gene shuffling, and marker-assisted selection to improve yields and resistance to stresses such as drought, salinity, and elevated temperatures (OECD 2009). In general, agriculture is a sector that also benefits from the development of the bioeconomy because certain agricultural varieties, such as corn and soy, constitute resources to produce biofuels and other bioproducts. However, it is also a sector that benefits from the bioeconomy since its introduction into agriculture leads to increased agricultural productivity.

The bioeconomy in the forestry sector refers to the use of forests to create products and services that help replace raw materials, products, and services based on fossil resources. In forestry, the bioeconomy connects the entire forest value chain, from the management and use of natural resources to the delivery of products and services. In addition, the bioeconomy increases the demand for forest goods and services, thus indirectly increasing economic opportunities for the entire sector, including bioenergy, wood construction, packaging products, chemicals, textiles, and more (Wolfslehner *et al.* 2016). Forestry is a vital sector in the bioeconomy since lignin and cellulose are obtained from wood; these are used to produce vanilla, plastic, oil, wax, asphalt, chemical products such as cellophane, sponges, ethanol, pastels, as well as paper, paper tissues, fibreboards, and the like (Forestry Extension Institute 2017). Just as biotechnology represents a part of the bioeconomy that provides technology for its development, forestry constitutes a part of the bioeconomy that provides resources (i.e., inputs for the development of the bioeconomy).

Like agriculture and forestry, fisheries, as part of the primary sector in the bioeconomy, play a role in providing resources for the secondary sector (i.e., the manufacture of bioeconomy-based products, which is described later in this chapter). The bioeconomy in the fisheries sector refers to fisheries management, aquaculture development, and the use of blue biotechnologies to manufacture food and non-food products (Čož-Rakovac and Topić Popović 2018). The so-called blue bioeconomy, which uses microorganisms, algae, and invertebrates as water biomass, is linked to the fisheries and aquaculture sector. This aquatic biomass is converted into food, feed, nutraceuticals, pharmaceuticals, cosmetics, energy, clothing, and more (European Commission 2021). As part of the blue bioeconomy, blue biotechnology represents the application of science and technology to living aquatic organisms to produce knowledge, goods, and services (OECD 2016).

In the secondary sector, the bioeconomy uses resources from the previously described primary economic activities for production. The importance of the bioeconomy in the food, beverage, and tobacco manufacturing sector can already be seen from various definitions that agree that bioeconomy is a concept of using natural resources and energy sources in the production of, among other things, food. The bioeconomy in the food, beverage, and tobacco manufacturing sector primarily uses clean energy from renewable sources in the production process, which is not a characteristic of this sector alone but of all secondary economic activity sectors. The first goal of the bioeconomy is to ensure food and food security, especially when considering population growth, changing dietary habits, and changing consumption patterns. The bioeconomy in the food, beverages, and tobacco manufacturing sector refers primarily to converting organic waste, food residues, and food waste into valuable and safe bio-based products (European Commission 2018). Put simply, the bioeconomy in the food, beverages, and tobacco manufacturing sector refers to the recovery and reuse of by-products and waste products to produce high-value-added molecules, new products, and energy recovery systems. In addition, the bioeconomy offers possibilities of creating new methods of food processing, as well as innovative techniques for food preservation to increase the overall sustainability of the supply chain of agri-food products and reduce waste.

The bioeconomy in the textile sector stemmed from the shortage of land for the expansion of cotton production, environmental concerns related to cotton and oil-based textile materials, and the development of new technologies for extracting fibres from wood (Maarit and Kallio 2021). In addition to using wood-based fibres, enzymes produced by industrial microorganisms, cellulose fibres extracted with nanotechnology from citrus waste, milk protein casein, and coffee grounds can also be used for textile production (Griestop *et al.* 2016).

The bioeconomy in the manufacture of wood products and furniture refers primarily to the use of wood and raw materials. The wood-based bioeconomy has the potential to contribute to the transition from a fossil to a sustainable circular economy. The bioeconomy products based on wood include furniture, sawn wood, wood panels, and also paper, textiles, biofuels, and pharmaceuticals made from wood fibre (Supin and Dzian 2018).

Paper manufacturing is a sector at the heart of the bioeconomy because it produces original wood-based products as raw materials and paper-based products and produces products that replace fossil fuel–based products. An example confirming the contribution of the paper manufacturing sector to the circular bioeconomy and the circular economy in general refers to the use of paper residues for renewable energy production, thereby reducing CO_2 emissions. In addition, residues from recycling can be converted into other valuable products, thus reaffirming the paper industry's contribution to the circular economy (Confederation of European Paper Industries 2013).

Biochemicals are chemical products wholly or partly derived from materials of biological origin, including biomass, raw materials, plants, algae, trees, marine organisms, and biological waste. The advantage of biochemicals compared to traditional chemicals is their eco-friendliness (European Commission 2018). For example, traditional chemicals used in manufacturing plastics, dyes, cosmetics, and fertilisers are made from petroleum and are not sustainable. Conversely, biochemicals used to manufacture chemical-based products, pharmaceuticals, plastics, and rubber are produced from biomass containing a complex mixture of substances; carbohydrates, fats, oils, and proteins are used to produce chemicals employing biotechnological processes. The production of biochemicals used to manufacture chemical-based products, pharmaceuticals, plastics, and rubber, is a valuable segment in the bioeconomy based on the potential of highly promising technologies, products, and markets (BMEL 2014).

The bioeconomy is mainly associated with biofuels and bioenergy; the production of biofuels and bioelectric power generation are two sectors that fully support the principles of the bioeconomy. Unlike traditional fossil fuels, which are formed by prolonged geological processes, biofuel is produced from biomass. The advantages of biofuels over fossil fuels include increasing national energy independence and reducing GHGs (Pearson and Turner 2014). Biofuel production is based on agricultural crops, including conventional food crops and specific energy crops, forestry, agricultural products, fishery or municipal waste products, and by-products and waste from the agricultural and food industries. The two most common types of liquid biofuels are bioethanol and biodiesel. Bioethanol is alcohol obtained by fermenting carbohydrates such as corn or sugar cane. Bioethanol can also be produced from cellulosic biomass obtained from non-food products such as trees and grass. Biodiesel is produced from oils or fats by transesterification (Islam *et al.* 2019).

Biofuels from biomass are used to produce bioenergy, which refers to all types of renewable energy obtained from biological raw materials known as biomass (Dahiya 2015). The benefits of bioenergy use include achieving renewable energy systems with low carbon content, the sequestration of atmospheric carbon, and several environmental and socio-economic benefits, including contributing to global climate change goals and environmental, social, economic, and sustainable goals (Röder and Welfle 2019). Furthermore, bioenergy represents the most extensive renewable contribution to the transport and heating sector and provides an essential share in electricity generation. However, bioenergy is facing significant challenges in electricity generation primarily related to high electricity generation costs and a limited cost-reduction area given that electricity generation is currently based on conventional power plant technologies, such as direct combustion of solid fuels (Belyakov 2019). Nevertheless, it bears stressing that, in addition to bioenergy from biomass, the bioelectricity generation sector also includes the use of renewable energy sources such as water and wind energy.

Global trends in the field of bioeconomy, revenues/expenditures, impact on GDP, impact on employment

The concept of bioeconomy is becoming increasingly important in the world; countries are developing strategies for the development of the bioeconomy and recognising global problems such as population growth, changing consumer habits, environmental issues, and climate change threats. When combined, all these problems result in unsustainability. The bioeconomy is recognised as one of the tools in the fight against unsustainability, which is highlighted by the fact that more than 50 countries around the world have dedicated bioeconomy strategies (OECD 2018).

Because the bioeconomy is a specific concept that encompasses parts of individual sectors, it is difficult to measure its contribution to global GDP, especially given the differences between countries and how they measure the contribution of the bioeconomy. For example, in the EU, the bioeconomy includes the sectors of agriculture, chemical industry, biofuels and bioenergy, animal feed, fisheries, food and beverage manufacturing, forestry, pharmaceuticals, pulp and paper, and textiles. The bioeconomy in the United States includes agriculture, the chemical industry, biorefining, forestry, and textiles. The US bioeconomy includes neither biofuels and bioenergy nor animal feed, fisheries, food and beverage manufacturing, the textile industry, and pulp and paper manufacturing. In Australia, the bioeconomy does not include pulp and paper production or the textile industry. The lack of a clear and homogeneous definition of the sectors included in the bioeconomy makes it impossible to compare or analyse the overall contribution of the bioeconomy to the global economy (FAO 2018). However, available indicators at the level of the EU, the United States, Asia, Australia, Africa, and Russia are presented next.

The bioeconomy represents a crucial part of the EU economy, having generated 4.7% of GDP and employed 8.9% of the EU-27 labour force (excluding the United Kingdom) in 2017. In 2017, the bioeconomy employed 17.5 million people at the EU level, of which the most (53%) referred to the agricultural sector and the least (only 0.1%) to the manufacture of liquid biofuels. Employment in the EU bioeconomy by sectors in 2017 is shown in Figure 1.1.

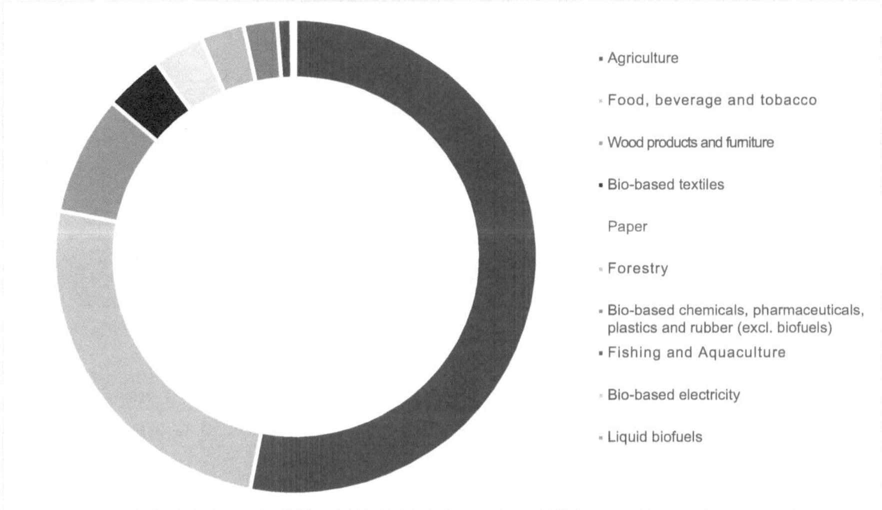

Figure 1.1 Employment in the bioeconomy by sectors of the EU-27 in 2017

Source: Author's compilation based on data available at: https://datam.jrc.ec.europa.eu/datam/mashup/BIOECONOMICS/#

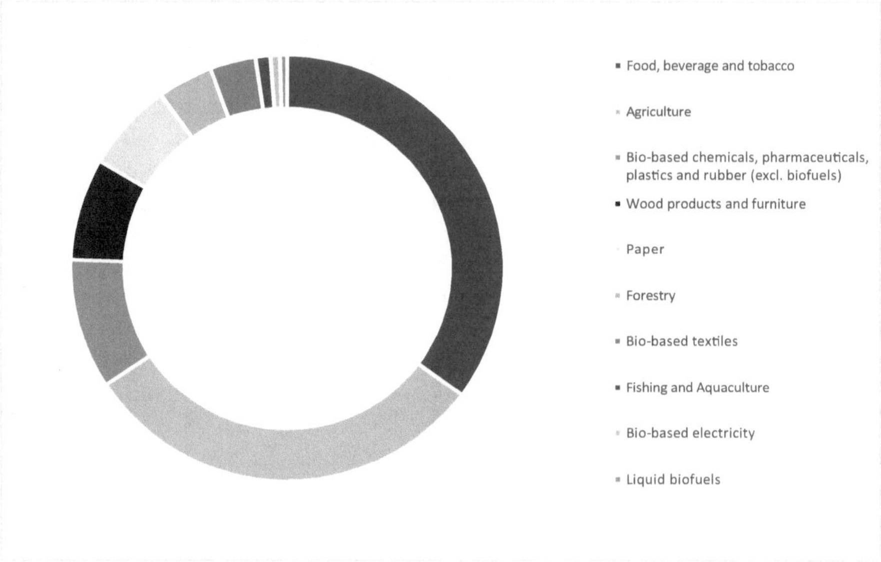

Figure 1.2 Added value in the bioeconomy by sectors in the EU-27 in 2017

Source: Author's compilation based on data available at: https://datam.jrc.ec.europa.eu/datam/mashup/BIOECONOMICS/#

In 2017, the total value added of the EU-27 bioeconomy amounted to EUR 614 billion; the highest value added was generated by the food, beverages, and tobacco industry (EUR 215 billion or 35.1%), and the lowest by liquid biofuels (only 0.5%). Value added by sectors of the EU-27 in 2017 is shown in Figure 1.2.

In 2016, the bioeconomy accounted for about 5.1% of the US GDP, or $959.2 billion, with a growth potential of up to 7.4% of GDP if available bio-based processes completely supplanted traditional non-biological processes (National Academies of Sciences, Engineering, and Medicine 2020). In 2016, the US bioeconomy employed **1.68 million people** directly and **2.98 million** indirectly. Compared to 2013, this represents an increase of 10.5% in direct employment and 19.2% in indirect employment (Figure 1.3).

In 2017, the total direct value added of the US bioeconomy amounted to $157 billion, and the indirect value added amounted to $302 billion. Compared to 2013, the direct value added increased by 24.6%, and the indirect value by 23.8% (Figure 1.4).

The bioeconomy in Asia is primarily developed in the countries of East and Southeast Asia, and the leading country in the development of the bioeconomy in Asia is Thailand. Thailand's bioeconomy is expected to account for

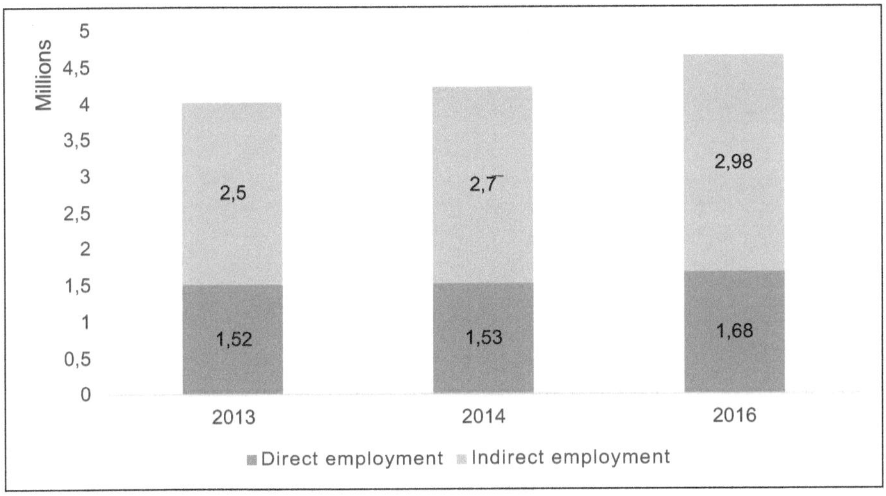

Figure 1.3 Direct and indirect employment in the US bioeconomy in 2013, 2014, and 2016 (millions)

Source: Author's compilation based on data available at: https://nap.nationalacademies.org/read/25525/chapter/6#85

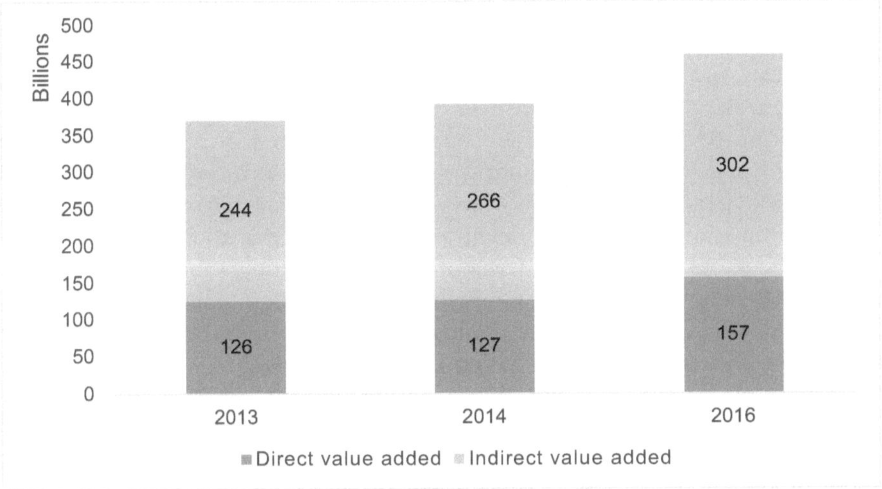

Figure 1.4 Direct and indirect value added in the US bioeconomy in 2013, 2014, and 2016 (millions)

Source: Author's compilation based on data available at: https://nap.nationalacademies.org/read/25525/chapter/6#85

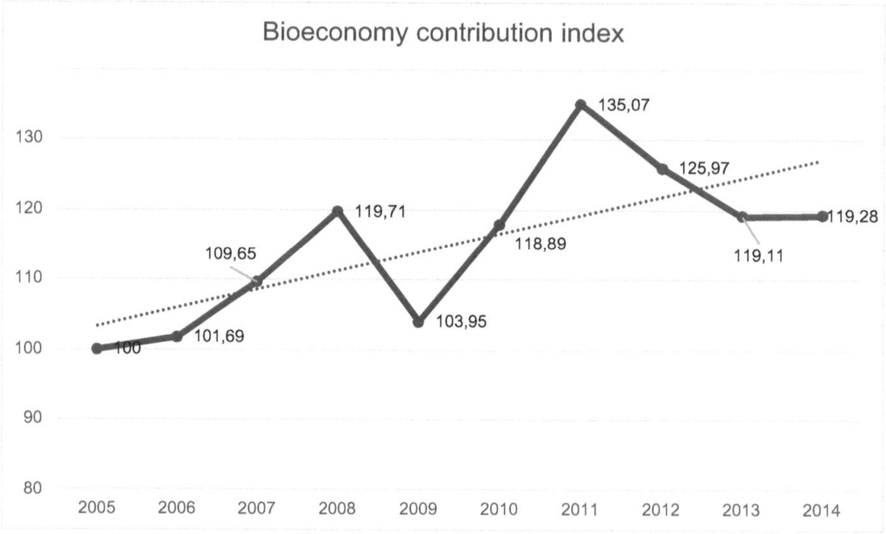

Figure 1.5 The Bioeconomy Contribution Index in Malaysia (2005–2015)

Source: Author's compilation based on data available at: www.bioeconomycorporation.my/bioeconomy-malaysia/investing-in-bioeconomy/bioeconomy-contribution-index/

2% of GDP by 2022 and 10% by 2037. The introduction of the bioeconomy in Thailand resulted in the growth of such sectors as biotechnology, bioenergy, biochemicals, and biopharmaceuticals, making Thailand the world's eighth and Asia's second-largest biofuels producer (Thailand Board of Investment).

In addition to Thailand, the bioeconomy is advanced in Malaysia; in 2015, it was estimated that the bioeconomy amounted to 11.3% of GDP, or MYR 131 billion (Malaysian ringgit). However, as the standard approach to measuring the economy against GDP lacks details regarding the progress and development of the bioeconomy, Malaysia has developed the Bioeconomy Contribution Index (FAO, 2018). The Bioeconomy Contribution Index combines five indicators: bioeconomy value added, productivity, investment, exports, and employment. The value of the Bioeconomy Contribution Index in 2015 is shown in Figure 1.5.

There is no strategy for developing the bioeconomy in Australia. Still, the bioeconomy is considered a driver of other sectors, such as agriculture, bio-based chemicals, biofuels, bioenergy, biorefining, fisheries, food and beverage manufacturing, forestry, and the pharmaceutical industry. Most Australian bioeconomy policies have been designed for research and development purposes. The lack of a specific bioeconomy development strategy (i.e., the lack of a structured bioeconomy) makes it impossible to systematically measure and monitor the contribution of the bioeconomy to the overall Australian economy (FAO 2018).

The Republic of South Africa represents the central point of bioeconomy development in Africa. In 2013, South Africa adopted the National Bioeconomy Strategy that addresses three key economic sectors: agriculture, industrial and environmental bio-innovation, and health. However, metrics that would measure the contribution of the bioeconomy to the economy of South Africa have not yet been implemented, although efforts have been made to create metrics to measure the contribution of the bioeconomy to economic sectors and innovation (FAO 2018).

Although Russia is globally recognised as a significant exporter of fossil raw materials and shows technological advances mainly in the defence industry, it is taking major steps to introduce the bioeconomy into its economy. In 2012, Russia adopted the BIO 2020 program as a tool for turning towards the bioeconomy. BIO 2020 is the first strategic document for improving the modern Russian bioindustry and includes the goals, priorities, and development agenda of the Russian biotechnology industry. It had nine areas, namely biopharmaceuticals, biomedicine, industrial, agricultural, food, forestry, environmental and marine biotechnology, and bioenergy. However, the BIO 2020 program has not become a full-fledged bioeconomy tool (Boyarov *et al.* 2020).

Future trends

The message conveyed throughout the chapter is that the bioeconomy is an extremely valuable tool in combating various current global issues, including population growth, poverty growth, environmental challenges and problems, climate change, and many more. As a concept based on the use of natural resources in production, the bioeconomy primarily tackles environmental issues but also indirectly deals with other social and economic issues.

The importance of the bioeconomy in the fight against current global issues has been recognised worldwide, which is why so many world powers have implemented strategies for the development of the bioeconomy. Smaller countries have followed their example; today, many nations around the world have their bioeconomy strategies. However, bioeconomy strategies worldwide differ according to a country's stage of development, its definition of the bioeconomy and sectors it encompasses, and the very desires and needs in shifting the economy from fossil to renewable sources.

The bioeconomy shows enormous potential in the energy, agricultural, food, and feed sectors. Furthermore, new products such as chemicals based on renewable energy sources show great promise concerning the transition from a fossil economy to a bioeconomy. In addition, new technologies such as genome editing in plant cultivation, metabolic engineering, and further digitisation of the primary sector are paving the way for furthering the development of the bioeconomy. The trust in bioeconomy-based innovation has been on the increase in view of the need to develop new bio-based products and optimise agriculture to feed a growing world population. The bioeconomy provides tools to address the challenge of increasing yields and optimising land use with the help of technological progress. Furthermore, the bioeconomy provides tools for achieving SDGs, especially SDG 9: industry, innovation, and infrastructure (Bioökonomierat 2018).

The future of the bioeconomy also depends on the application of biotechnology, which has the potential to improve and manage food, feed, and fibre crops, and is driven by growing demand and increased agronomic stress due to climate change. Furthermore, the expected long-term increase in the price of fossil fuels due to the decline in the supply of low-cost oil, as well as the increase in demand for energy, and restrictions on GHG production, constitute the basis for the growth of the biomass market, including non-food crops such as grass and wood, as raw materials for biofuels, chemicals, and plastics. The bioeconomic potential of biotechnology lies in the use of plants to produce valuable chemicals such as biopharmaceuticals and the production of nutraceuticals from plant and animal sources. Trends highlight the importance of the future role of biotechnology and the bioeconomy (OECD 2009).

The projected increase in demand for biomass for food and industrial purposes will lead to the need for a sustainable increase in agricultural, forestry, fisheries, and aquaculture capacities. On the other hand, the development of production systems with reduced GHG emissions will lead to the adaptation and mitigation of the harmful effects of climate change, such as droughts and floods, but also contribute to the transition to a low-carbon economy with efficient resources (European Commission 2012).

In short, to make use of the potential of the bioeconomy, it is necessary to manage natural resources wisely and sustainably and tackle the challenges of developing the bioeconomy, but also take regional and local aspects into account when preparing the bioeconomy strategies. Existing policies need to be carefully complemented, bearing in mind that the needs of less developed and developing countries are technologically different from the needs of industrialised, highly developed countries. Future efforts should certainly include educational, financial, and policy measures to promote a thriving bioeconomy market. Knowledge and funding are vital to improving the bioeconomy in the future. National strategies should prioritise sufficient funding and access to capital, as well as training programmes for professionals and the general population. It will also be necessary to pool knowledge between the industrialised and developing countries to ensure positive developments in the bioeconomy (Bioökonomierat 2018).

Bibliography

Beg, M., 2018. Održivi ekonomski razvoj – prepreke i rješenja. In G. Družić and I. Družić, eds. *Zbornik radova znanstvenog skupa: Modeli razvoja hrvatskog gospodarstva*. Zagreb: Faculty of Economics and Business, University of Zagreb, 371–394.

Beluhova-Uzunova, R., Shishkova, M., and Ivanova, B., 2019. Concepts and key sectors of the bioeconomy. *Trakia Journal of Sciences*, 227–233.

Belyakov, N., 2019. Bioenergy. In N. Belyakov, ed. *Sustainable Power Generation: Current Status, Future Challenges, and Perspectives*. London: Academic Press, an imprint of Elsevier, 461–474.

Bioökonomierat, 2018. *Future Opportunities and Developments in the Bioeconomy – A Global Expert Survey*. Berlin: Bioökonomierat.

Birner, R., 2018. Bioeconomy concepts. In I. Lewandowski, ed. *Bioeconomy*. Cham: Springer International Publishing, 17–38.

BMEL, 2014. *National Policy Strategy on Bioeconomy: Renewable Resources and Biotechnological Processes as a Basis for Food, Industry and Energy*. Berlin: Federal Ministry of Food and Agriculture.

Bogdana, A., Istudorb, N., Gruiac, R., Tobăa, G., Bulza, N., Gâf-Deacd, I., Chelmu, S., Găvan, C., Prică, I., and Paşalăuf, C., 2014. New holistic approach of bioeconomics and ecoeconomics theories, practical bridging from the green economy to blue economy, through new integrated and innovative paradigm about "Bio-eco-geo-economy". *Procedia Economics and Finance*, 83–90.

Boyarov, A., Osmakova, A., and Popov, V., 2020. Bioeconomy in Russia: Today and tomorrow. *New Biotechnology*, 60, 36–43. Available from: https://doi.org/10.1016/j.nbt.2020.08.003

Bugge, M. M., Hansen, T., and Klitkou, A., 2016. What is the bioeconomy? A review of the literature. *Sustainability*, 8, 691. Available from: https://doi.org/10.3390/su8070691

Carus, M., 2017. The bioeconomy is much more than a circular economy. *Blickwinkel*, August. Available from: https://www.brain-biotech.com/blickwinkel/circular/the-bioeconomy-is-much-more-than-a-circular-economy/

Confederation of European Paper Industries, 2013. *Sustainability Report*. CEPI, Brussels. https://www.cepi.org/cepi-sustainability-report-2013/

Cotula, L., Dyer, N., and Vermeulen, S., 2008. *Fuelling Exclusion? The Biofuels Boom and Poor People's Access to Land*. London: IIED.

Čož-Rakovac, R., and Topić Popović, N., 2018. Bioeconomy in fisheries and aquaculture. In *Book of Abstracts 9th International Congress of Food Technologists, Biotechnologists and Nutritionists*. Zagreb: Croatian Society of Food Technologists, Biotechnologists and Nutritionists, 46–46.

Dahiya, A., 2015. Introduction. In A. Dahiya, ed. *Bioenergy: Biomass to Biofuels*. London: Academic Press, an imprint of Elsevier, 1–4.

D'Amato, D., Droste, N., Allen, B., Kettunen, M., Lähtinen, K., Korhonen, J., Leskinen, P., Matthies, B. D., and Toppinen, A., 2017. Green, circular, bio economy: A comparative analysis of sustainability avenues. *Journal of Cleaner Production*, 716–734.

D'Amato, D., Veijonaho, S., and Toppinen, A., 2020. Towards sustainability? Forest-based circular bioeconomy business models in Finnish SMEs. *Forest Policy and Economics*, 110, 101848. Available from: https://doi.org/10.1016/j.forpol.2018.12.004

Enriquez, J., 1998. Genomics and the world's economy. *Science*, 925–926.

European Commission, 2012. *Innovating for Sustainable Growth: A Bioeconomy for Europe*. Brussels: European Commission.

European Commission, 2013. Bioeconomy. Available from: https://ec.europa.eu/programmes/horizon2020/en/h2020-section/bioeconomy

European Commission, 2018a. *A Sustainable Bioeconomy for Europe: Strengthening the Connection between Economy, Society and the Environment*. Luxembourg: Publications Office of the European Union.

European Commission, 2018b. *A Sustainable Bioeconomy for Europe: Strengthening the Connection between Economy, Society and the Environment: Updated Bioeconomy Strategy*. Luxembourg: Publications Office of the European Union.

European Commission, 2018c. In-depth Analysis in Support of the Commission Communication COM(2018) 773. A European Long-term Strategic Vision for a Prosperous, Modern, Competitive and Climate Neutral Economy. Bruxelles: European Commission.

European Commission, 2021. Blue bioeconomy and blue biotechnology, 3 September. Available from: https://ec.europa.eu/oceans-and-fisheries/ocean/blue-economy/blue-bioeconomy-and-blue-biotechnology_en

European Investment Bank, 2018. *Agriculture and Bioeconomy: Unlocking Production Potential in a Sustainable and Resource-efficient Way*. Luxembourg: EIB.

FAO, 2016. *How Sustainability Is Addressed in Official Bioeconomy Strategies at International, National and Regional Levels: An Overview*. Rome: FAO.

FAO, 2018. *Assessing the Contribution of Bioeconomy to Countries' Economy: A Brief Review of National Frameworks*. Rome: Food and Agriculture Organization of the United Nations.

Fargione, J., Hill, J., Tilman, D., Polasky, S., and Hawthrone, P., 2008. Land clearing and the biofuel carbon debt. *Science*, 1235–1237.

Forestry Extension Institute, 2017. *Bioeconomy and European Forest Week 2017*. Warsaw: Forestry Extension Institute – Norway.

Geissdoerfer, M., Savaget, P., Bocken, N., and Hultink, E., 2017. The circular economy – A new sustainability paradigm? *Journal of Cleaner Production*, 757–768.

Gheewala, S. H., Berndes, G., and Jewitt, G., 2011. The bioenergy and water nexus. *Biofuels, Bioproducts and Biorefining*, 353–360.

Giampietro, M., 2019. On the circular bioeconomy and decoupling: Implications for sustainable growth. *Ecological Economics*, 143–156.

Griestop, L., Colthorpe, J., and Wirsching, S., 2016. *Bioeconomy in Everyday Life*. Berlin: Biocom AG.

Hiemann, T., 2019. Bioeconomy and SDGs: Does the bioeconomy support the achievement of the SDGs? *Earth's Future*, 43–57.

International Advisory Council on Global Bioeconomy, 2020. *Global Bioeconomy Policy Report (IV): A Decade of Bioeconomy Policy Development Around the World*. Berlin: International Advisory Council on Global Bioeconomy.

Islam, M., Hasanuzzaman, M., Pandey, A., and Rahim, A., 2019. Modern energy conversion technologies. In M. Hasanuzzaman and A. Rahim, eds. *Energy for Sustainable Development: Demand, Supply, Conversion and Management*. London: Academic Press, an imprint of Elsevier, 19–39.

Krugman, P., and Wells, R., 2012. *Economics*. New York: Worth Publishers.

Maarit, A., and Kallio, I., 2021. Wood-based textile fibre market as part of the global forest-based bioeconomy. *Forest Policy and Economics*, 13, 102364. Available from: https://doi.org/10.1016/j.forpol.2020.102364

Nagarajan, D., Lee, D., and Chang, J., 2021. Circular bioeconomy: An introduction. In A. Pandey, R. Tyagi, and S. Varjani, eds. *Biomass, Biofuels, Biochemicals: Circular Bioeconomy – Current Developments and Future Outlook*. (s.l.): Elsevier, 3–23.

National Academies of Sciences, Engineering, and Medicine, 2020. *Safeguarding the Bioeconomy*. Washington, DC: The National Academies Press.

OECD, 2008. *Biofuel Support Policies: An Economic Assessment*. Paris: OECD.

OECD, 2009. *The Bioeconomy to 2030: Designing a Policy Agenda. Organisation for Economic Co-operation and Development*. Paris: OECD.

OECD, 2013. Biotechnology. In OECD, ed. *OECD Factbook 2013: Economic, Environmental and Social Statistics*. Paris: OECD Publishing, 156–157.

OECD, 2016. *The Ocean Economy in 2030*. Paris: OECD Publishing.

OECD, 2018. *Meeting Policy Challenges for a Sustainable Bioeconomy*. Paris: OECD Publishing.

Pearson, R., and Turner, J., 2014. Improving the use of liquid biofuels in internal combustion engines. In K. Waldron, ed. *Advances in Biorefineries: Biomass and Waste Supply Chain Exploitation*. Sawston: Woodhead Publishing, 389–440.

Pietzsch, J., and Schurr, U., 2017. Introduction. In J. Pietzsch, ed. *Bioeconomy for Begginers*. Berlin: Springer-Verlag, 1–10.

Purnhagen, K., and Matthews, A., 2019. European agriculture and the bioeconomy: A historical overview. In *EU Bioeconomy Economics and Policies: Volume I*. Cham: Palgrave Macmillan, 27–35.

Röder, M., and Welfle, A., 2019. Bioenergy. In T. Letcher, ed. *Managing Global Warming: An Interface of Technology and Human Issues*. London: Academic Presss, an imprint of Elsevier, 379–398.

Rogers, J., Stokes, B., Dunn, J., Cai, H., Wu, M., Haq, Z., and Baumes, H., 2017. An assessment of the potential products and economic and environmental impacts resulting from a billion ton bioeconomy. *Biofuels, Bioproducts, and Biorefining*, 110–128.

Ronzon, T., Piotrowski, S., M'barek, R., Carus, M., and Tamošiūnas, S., 2020. *Jobs and Wealth in the EU Bioeconomy*. Brussels: European Commission Joint Research Centre (JRC).

Supin, M., and Dzian, M., 2018. Influence of bio-economy on the development of wood. In *11th International Scientific Conference Wood, EMA 2018 Proceedings of Scientific Papers*. Belgrade: University of Belgrade – Faculty of Forestry, 30–37.

Thailand Board of Investment, n.d. *Thailand's Bioeconomy*. Bangkok: Thailand Board of Investment.

UNEP, 2011. *Towards a Green Economy: Pathways to Sustainable Development and Poverty Eradication*. Nairobi: UNEP.

United Nations, 1987. *Report of the World Commission on Environment and Development: Our Common Future*. New York: United Nations; World Commission on Environment and Development.

Viaggi, D., 2021. Exploring the economics of the circular bioeconomy. In V. Venkatramanan, S. Shah, and R. Prasad, eds. *Sustainable Bioeconomy Pathways to Sustainable Development Goals*. Singapore: Springer, 1–10.

Vogelpohl, T., and Töller, A., 2021. Perspectives on the bioeconomy as an emerging policy field. *Journal of Environmental Policy & Planning*, 143–151.

Wolfslehner, B., Linser, S., Pülzl, H., Bastrup-Birk, A., Camia, A., and Marchetti, M., 2016. *Forest Bioeconomy – A New Scope for Sustainability Indicators*. Joensuu, Finland: European Forest Institute.

2 Bioeconomy policies across the globe

Marina Funduk

Introduction

In recent years, the concept of the bioeconomy has become more visible in the policy process worldwide. As a result, more and more countries are developing strategies and action plans dedicated to the bioeconomy. From the first decade of the 21st century, almost 60 countries around the world adopted their bioeconomy-related policies.

In Europe, the European Union (EU) is seen as one of the political pioneers of the bioeconomy. The support for bioeconomy development goes back to 2005 when its Commissioner for Research, Science and Innovation presented the Knowledge-Based Bio-Economy concept, adopted under the German EU presidency in 2007. The experience accumulated over the years and with new societal expectations triggered the launch of an EU Strategy on Bioeconomy in 2012 (Patermanna and Aguilar 2018). Currently, nine EU member states (Austria, Finland, France, Germany, Ireland, Italy, Latvia, Spain, and the Netherlands) have a national bioeconomy strategy. Other member states are involved in national, regional (e.g., Flanders) (Flemish Government, Department of Economy, Science and Innovation 2020), or macro-regional (e.g., Nordic) (Nordic Council of Ministers 2018) bioeconomy development (European Commission, Joint Research Centre 2018).

A significant stimulus to the development of national and regional bioeconomy strategies was provided by the Organisation for Economic Co-operation and Development (OECD) in 2009 with its strategy paper entitled *The Bioeconomy to 2030: Designing a Policy Agenda* (OECD 2009). This publication can be seen as a starting point for understanding strategies and policy formulation on the bioeconomy. It emphasises the progress in biological sciences, which can now offer solutions for many health- and resource-related issues the world is facing. These solutions include the development of new technologies, which provide a motor for increased sustainability in economies worldwide. However, defining a policy agenda is crucial for the implementation of these research findings and innovations (Staffas *et al.* 2013).

Many countries are still in the process of building their bioeconomy strategies, while a number of policies are already in place for individual bioeconomy sectors.

DOI: 10.4324/9781003223733-3

This chapter is based on the report by the International Advisory Council on Global Bioeconomy entitled the *Global Bioeconomy Policy Report (IV): A Decade of Bioeconomy Policy Development Around the World* and provides an overview of the 19 dedicated national and macro-regional bioeconomy policy strategies published across the globe (Austria, Costa Rica, the EU, Finland, France, Germany, Ireland, Italy, Japan, Latvia, Malaysia, Nordic countries (involving Denmark, Finland, Iceland, Norway, Sweden, the Faroe Islands, Greenland, and Åland), Norway, South Africa, Spain, Thailand, the United Kingdom, the United States, East Africa).

The German Federal Government was one of the first governments in the world to put the bioeconomy on its research policy agenda (Schütte 2018). It adopted the *National Research Strategy BioEconomy 2030: Our Route towards a Biobased Economy* (German Federal Ministry of Education and Research 2011) in 2011, while the dedicated *National Policy Strategy on BioEconomy* (German Federal Ministry of Food and Agriculture 2014) was published three years later. The German Bioeconomy Council is the most responsible for this pioneering work on the bioeconomy in Germany. Established in 2009 as an independent advisory board to the government for all matters regarding the bioeconomy, it provided recommendations that the national strategy itself is based on (Staffas *et al.* 2013). The German Bioeconomy Council has developed three key overviews of bioeconomy policies in recent years (German Bioeconomy Council 2015a, 2015b, 2018). One focuses on G7 countries, another on the situation worldwide, and the third provides an update on developed strategies until 2018. In 2020, the International Advisory Council on Global Bioeconomy issued the most recent, fourth report on bioeconomy policies around the world.

In 2012, the European Commission launched the EU Strategy on Bioeconomy – *Innovating for Sustainable Growth: A Bioeconomy for Europe* (European Commission 2012), while the United States published its *National Bioeconomy Blueprint* (The White House 2012) in the same year. The EU Bioeconomy Strategy in 2012 provided a decisive push for the development of national bioeconomy strategies in Europe. It defined the bioeconomy as "the production of renewable biological resources and the conversion of these resources and waste streams into value-added products" (European Commission 2012, p. 3). The strategy advocated for higher production with renewable biological resources and "encouraged cascading uses of biomass, bio-based products, and bio-based waste streams along pre-existing and novel value chains" (Ronzon *et al.* 2022). Additionally, the strategy supported research and innovation targeted at endorsing the development of new bio-based products aimed at realising a transition towards a low-carbon economy with offering new market opportunities to farmers, foresters, and fishers as biomass suppliers.

The bioeconomy in Europe has, since 2012, gradually strengthened its integrity as a green growth strategy by broadening its sphere to embrace "related services activities and by integrating the notion of environmental preservation" (Ronzon *et al.* 2022). This more comprehensive conceptualisation is reflected in the EU's revised definition of the bioeconomy, where it

> includes and interlinks land and marine ecosystems and the services they provide, all primary production sectors that use and produce biological resources . . . and all economic and industrial sectors that use biological resources and processes to produce food, feed, bio-based products, energy and services.
>
> (European Commission 2018a)

In its revised definition, the bioeconomy has become essential in recent EU policies, like the implementation of the *Circular Economy Action Plan* (European Commission 2020a), the forestry strategy, the blue (bio)economy, the *Common Agricultural Policy*, and the *European Green Deal* (European Commission 2019).

The EU Bioeconomy Strategy from 2018 *A Sustainable Bioeconomy for Europe: Strengthening the Connection between Economy, Society and the Environment. Updated Bioeconomy Strategy* (European Commission 2018a) emphasises that the European bioeconomy needs to have "sustainability and circularity at its heart" to be successful. It must be both sustainable and circular. The strategy aims to accelerate the deployment of a sustainable European bioeconomy and to maximise its contribution towards the 2030 Agenda (UN 2015), Sustainable Development Goals (SDGs) and the Paris Agreement (UN 2016). Challenges of climate change together with the reduction of CO_2 are "powerful global drivers toward more efficient use of biological resources" (Lange *et al.* 2021). The bioeconomy contributes both to climate change mitigation and to fulfilling the SDGs (Lange *et al.* 2021; Philippidis *et al.* 2018; Heimann 2019; M'barek *et al.* 2019). An important step in the relationship between the SDGs and the bioeconomy was the adoption of the 2030 Agenda in 2015. Since then, that relationship has solidified in many dedicated policy strategies around the world.

A noticeable trend in newer bioeconomy strategies refers to the heightened role of the circular economy. However, there are different views. If we look at the countries with dedicated bioeconomy strategies, we can see that on the one hand, the EU and Italy view the bioeconomy as a subsector of the circular economy. The strategy in Italy refers to a transition towards a circular bioeconomy where the "production and use of renewable bioresources and their conversion into value-added products is part of a circular system" (International Advisory Council on Global Bioeconomy 2020b, p. 155). Likewise, France, Latvia, and Norway view the bioeconomy as an essential part of the circular economy and the United Kingdom as an opportunity for developing a more circular economy. On the other hand, the view that the bioeconomy is circular by nature has gained in force. Strategies in Austria, Costa Rica, Germany, Japan, Malaysia, and the Nordic countries view circular concepts as essential elements of the bioeconomy. Similarly, the bioeconomy in Ireland promotes circularity through innovations that reuse and recycle materials and maximise resource efficiency.

Besides developing the bioeconomy strategies and action plans, countries gather in Global Bioeconomy Summits. The first Global Bioeconomy Summit was organised in 2015 in Berlin, where the essential contribution of a sustainable bioeconomy to achieving the UN SDGs was highlighted. The second one was organised in 2018, where it was stressed that no one solution fits all and that the political strategies point to different opportunities and pursue key objectives depending on their prevailing industrial and economic profiles and natural resource potential (International Advisory Council on Global Bioeconomy 2020b, p. 7). The Global Bioeconomy Summit in 2018 encouraged "innovations and international collaboration" as vital elements of an inclusive bioeconomy (International Advisory Council on Global Bioeconomy 2020a, p. 5). When it was organised in 2020, the third Global Bioeconomy Summit emphasised that the bioeconomy has achieved significant progress in recent years in moving us towards a new economy.

It has been increasingly recognised that the strength of a bioeconomy lies in its diversity. The importance of global cooperation for sustainability and the innovations needed to drive it are laid out in the *Communiqué of the Global Bioeconomy Summit 2020* (International Advisory Council on Global Bioeconomy 2020a). It emphasises utilising the potential of life sciences, digitisation, and their interlinkages; opportunities for bioeconomy jobs through partnerships and innovation; mobilisation of finance for bioeconomy development; increasing involvement of industry and business; resilient value chains; strengthening demand-side policy approaches; and a global platform.

In addition to national policy initiatives, there is the increasing engagement of macro-regional and international actors in bioeconomy development across the globe. In Europe, the EU Bioeconomy Strategy provided the initial push. In East Africa, national efforts to develop dedicated bioeconomy policy initiatives were stimulated by the publication of the *Regional Bioeconomy Strategy for Eastern Africa* (International Advisory Council on Global Bioeconomy 2020b, p. 13). The number of regional strategies in Europe and Latin America have increased. Similarly, new dynamics in bioeconomy policy development are evolving outside of governments, with stakeholder- and industry-driven strategies on the rise.

Core elements of bioeconomy policy strategies

Strategies generally provide a vision statement with several goals, objectives, and qualitative targets. Less frequently, quantitative targets are provided, focusing on increasing economic output, employment, or exports like in Finland, Italy, Latvia, Malaysia, and the United Kingdom (International Advisory Council on Global Bioeconomy 2020b, p. 156). Principles and guidelines that align with the SDGs are displayed in the strategies of Austria, Germany, Ireland, Japan, and the Nordic countries. However, according to the report from the International Advisory Council on Global Bioeconomy, the extent to which policy goals are translated into concrete measures has continued to be limited, with vaguely defined measures and budget appropriations.

The trend of developing compatible action plans to the bioeconomy strategies is on the rise. Nevertheless, only the EU, France, Ireland, Italy, Norway, and Spain have published their action plans, while Costa Rica, Germany, Japan, and the UK are in the process of creating them. Those action plans vary significantly from one another. Some have concrete policy measures with appropriate indicators and some milestones (EU); some have regular progress reports with identified leads, co-leads, and key consultative stakeholders (Ireland); and some are more general guiding documents for how main policy actors should best collaborate (Norway). The action plans are also formulated in diverse ways. Some of them have outsourced their development to public research and innovation corporations, like Norway. Some of them outsourced it to the consortia representing the government, industry, and research community (the United Kingdom) or to other multi-stakeholder bodies (Costa Rica) or to high-level inter-ministerial working groups (Italy, Ireland) (International Advisory Council on Global Bioeconomy 2020b, p. 156).

In most countries, major efforts have been made to involve stakeholders in the strategy development process. Standard engagement elements include a consultation process, stakeholder working groups to set up draft documents, public consultation, inter-agency efforts, and conferences. In addition, some countries and regions have already updated their dedicated bioeconomy policy strategy based on a comprehensive strategy review process, like the EU, Germany, Italy, and Japan, while others have not. In the United States and South Africa, the focus today is on regulating modern biotechnologies and not on the development of policy strategies.

Bioeconomy strategy goals

Almost 60 countries worldwide have shaped their bioeconomy-related policies and strategies. Nevertheless, the definition of which areas belong to the bioeconomy and what objectives countries strive for along their bioeconomic paths are very different (Pietzsch and Schurr 2017). Some countries have formulated comprehensive bioeconomic strategies that address regional problems and relate to global issues, while others are much more specific regarding their own country's concrete objectives.

The source of political motivation for promoting bioeconomy development varies according to a "country's raw material base, economic specialisation, and the level and path of development" (German Bioeconomy Council 2015a, 2015b). Different national motivations to pursue bioeconomy strategies can be divided into four categories from an economic perspective. Some countries have a problem with structural food shortages. The bioeconomy is therefore primarily seen as a way to enable more effective production of food and feed, which is basically focused at food security. This is the case in one of the East African countries, in Tanzania. The Tanzanian *National Biotechnology Policy* of 2010 focuses on the agricultural sector. Its main goal is to promote food security and ensure that the country is self-sufficient and not dependent on expensive imports (Pietzsch and Schurr 2017). Other countries in Africa such as Kenya, Mozambique, and Uganda have a similarly strong focus on food security.

Some countries have a large and almost inexhaustible quantity of natural resources, which they have not yet learnt to use efficiently enough. Consequently, the bioeconomy is understood as "a way to establish new biomass value chains that profitably link the raw materials of the primary sector with downstream sectors to increase the gross national product" (Pietzsch and Schurr 2017). This is the case in one of the countries in Europe, Finland. Forests cover almost three-quarters of the country's surface area. As a consequence, forestry is at the core of the 2014 *Finnish Bioeconomy Strategy – Sustainable Growth from Bioeconomy* (Albrecht and Ettling 2014). Strengthening the timber market and diversifying the supply of timber products contributes to Finland's prosperity and competitiveness (Pietzsch and Schurr 2017). Argentina, Brazil, and Russia are among the countries that have an abundant supply of biomass, the starting point for their bioeconomic strategies.

Some countries are home to diverse and large industries that depend on sufficient amounts of raw materials (Pietzsch and Schurr 2017). Accordingly, they are interested in new sources of raw materials. This is the case in Germany. The *National Research Strategy Bioeconomy* of 2011 and the *National Policy Strategy Bioeconomy* of 2014 are comprehensively dedicated to the bioeconomy. Also, Germany updated their bioeconomy strategy in 2020. Countries that are comparable to Germany are Japan and the United States.

Some countries are on the verge of becoming industrial countries with their abundant natural resources and considerable high-tech potential. Therefore, the use of technology in the bioeconomy is seen as a means to cross the threshold. This is the case in Malaysia which is rich in soil resources and biomass. Moreover, it has experienced a rapid economic rise in recent years. Malaysia regards biotechnology as a key driver of its future growth. *National Biotechnology Policy* from 2005 emphasised Malaysia's goal to become a knowledge-based economy by 2020. In this context, the document *Bioeconomy Transformation Programs* was published in 2012, followed by the amended *National Biomass Strategy 2020* in 2013. The strategy focuses on the agricultural biomass (palm oil) and concentrates on promoting technologies that generate a higher added value from the country's biological resources. Thailand, South Africa, and India have comparable starting positions.

In its *Global Bioeconomy Policy Report (IV): A Decade of Bioeconomy Policy Development Around the World*, the International Advisory Council on Global Bioeconomy points out that, in 2015, oil-importing countries with considerable biomass resources often strived for higher energy independence and sought to increase the added value of their biological resources. Industrialising countries with a significant share of the rural population and primary industry jobs also considered bioeconomy development as a means of fostering rural development and social inclusion. In contrast, industrialised countries with fewer biological resources focused more on the opportunities from the industrialisation of biology and on creating added value from biosciences.

More recent strategies consider a wider range of goals due to recognising the increasing complexity of the bioeconomy and the integration of topics, such as sustainability, climate, and a circular economy. Also, there is no longer a clear hierarchy of goals, but rather a more diverse set of equal goals.

Bioeconomy strategies from Austria, Italy, Japan, Spain, and Thailand address global societal challenges. In this view, the bioeconomy is recognised as a strategy for coping with climate change, reducing greenhouse gases, promoting the decarbonisation of production and consumption processes, and contributing to the Paris Agreement (International Advisory Council on Global Bioeconomy 2020b, p. 154). In general, almost all strategies emphasise the bioeconomy's contribution to the SDGs, either in the context of promoting a sustainable economy or within their overall sustainability policy. Focus is on transformation of both economy and society. The EU is pushing for a green transformation while Austria seeks a sustainable social transformation. Japan and Spain emphasise the importance of the bioeconomy's social dimension and the need for public engagement, while Austria, Germany, and the Nordic countries highlight the need for changes in behaviour and values of producers and consumers

The bioeconomy is increasingly viewed as a strategy for reindustrialisation. In the EU, Germany, Italy, and the United Kingdom, the bioeconomy is highlighted for its ability to transform former coal or structurally weak areas, contribute to the renewal of industry, or modernise the primary production sectors. Italy focuses strongly on converting abandoned land and industrial sites such as former chemical complexes.

Strategies that have been recently updated, such as those of the EU, Germany, Italy, and Japan, wish to better align with other policy priorities. One of the central aims of the updated EU strategy is to adapt to a policy context that has significantly changed, especially due to the EU circular economy and the Paris Climate Agreement and the 2030 Agenda for Sustainable Development. More recent strategies such as those of Ireland and the United Kingdom react to the uncertainties brought about by Brexit. In Japan, the bioeconomy strategy responds to global changes in international power structures, with the world's centre of economic power shifting from the existing industrialised countries of Europe, Japan, and the United States to developing countries in Asia and Africa (International Advisory Council on Global Bioeconomy 2020b, p. 155). It views the global trends of rapid population and economic growth in Asia and Africa as an opportunity for increasing demand for goods, such as healthier food varieties and medicines.

Strategies and action plans of Costa Rica, Italy, and Japan were published in the time of the global Covid-19 pandemic and accordingly recognise the potential of the bioeconomy to produce more efficiently, maintain value chains, and sustain jobs and livelihoods while ensuring the sustainable use and rehabilitation of nature. The Italian *Implementation Action Plan (2020–2025) for the Italian Bioeconomy Strategy BIT II* (National Bioeconomy Coordination Board of the Presidency of Council of Ministers 2021) emphasises the resiliency of the circular bioeconomy and shows that investments in circular economy can be seen as a catalyst of socio-economic post-Covid-19 restarting. A detailed action plan 2020–2025 includes actions clustered into four main macro areas (policy/standards, pilot actions, regeneration of ecosystem services, and stakeholder's engagement). However, the theme of the bioeconomy contributing to increased resilience plays a less significant role in strategies, with the exception of the Nordic countries. Japan's updated strategy published in 2020, *Bio-Strategy 2020* (Council for Integrated Innovation Strategy 2020), highlights the bioeconomy's role in the post-Covid-19 era in developing measures against a future public health crisis. Japan, Malaysia, South Africa, and the United States consider innovations in the bioeconomy as significant for improving human health like early disease detection and cheaper, accessible medicines (International Advisory Council on Global Bioeconomy 2020b, p. 155).

Prioritisation and specialisation in policy strategies

Bioeconomy strategies across the globe have different priorities and different focuses regarding their local conditions. Almost all countries look to make advancements in agriculture, forestry, and fisheries through "improved cultivation, harvesting and processing technologies, precision farming, utilisation of digitalisation, organic farming, urban and vertical farming" (International Advisory Council on Global Bioeconomy 2020b, p. 156). The forestry sector of the bioeconomy is prioritised in the strategies of Austria, Finland, Ireland, and Japan. Austria and Finland's strategy focuses on growth potential in "new and advanced materials based on cellulose, lignin, wood fibre, wood wool, and fibre plants" (International Advisory Council on Global Bioeconomy 2020b, p. 156). Austria, Costa Rica, Japan, and the Nordic countries focus on wooden construction.

The EU Action plan 2018, *Bioeconomy: The European Way to Use Our Natural Resources* (European Commission 2018b), emphasises three priorities: (1) strengthen and scale up the bio-based sectors, unlock investments and markets; (2) deploy local bioeconomies rapidly across the whole of Europe, and (3) understand the ecological boundaries of the bioeconomy.

In Austria, Finland, South Africa, and Spain, the bioeconomy is connected to the protection of water resources, whether by improving technologies for water efficiency and water recycling in Finland or by sensibly utilising unused nutrients in sludge from biogas or sewage plants in Austria.

A lot of dedicated bioeconomy strategies (Austria, Costa Rica, the EU, Finland, France, Germany, Ireland, Latvia, Nordic countries, Norway, Spain, Thailand, and the United Kingdom) focus on biorefinery development. The EU plans to develop a roadmap for the deployment of 300 biorefineries on a small and decentralised scale with a focus on sustainable chemicals. Italy sees biorefinery development as an opportunity for the reindustrialisation and decontamination of former oil refineries and chemical plants in rural areas. Other countries, such as Costa Rica, Latvia, Ireland, France, and Nordic countries see biorefinery development as "essential for rural areas to maintain a degree of self-sufficiency" (International Advisory Council on Global Bioeconomy 2020b, p. 157).

The EU, Germany, Ireland, and Latvia emphasise the potential of new products with totally new functions and the potential to create new markets. Austria, Japan, Malaysia, and the United Kingdom prioritise bio-based plastics while EU, Japan, and Malaysia call for plastic-free seas and oceans by targeting marine biodegradable plastics. France focuses on "protein production as a source of new and alternative food resources" (International Advisory Council on Global Bioeconomy 2020b, p. 157). Development of the bioeconomy is strongly related to the medical sector in Japan, Malaysia, and Thailand. Ireland and Norway have growing biopharmaceutical sectors. Ireland, Japan, and Thailand examine functional food markets while Malaysia and the Nordic countries promote innovations in food production systems.

Regarding the role of technology, the EU, Finland, and the United States emphasise the application of biotechnology in general while Japan, South Africa, and the United Kingdom focus on industrial biotech. Costa Rica, Germany, Thailand, the United Kingdom, and the United States concentrate on synthetic biology while the EU, Italy, Ireland, Latvia, Nordic countries, and Spain focus on marine biotech. All strategies emphasise the importance of converging technologies and digitalisation. Connecting bioscience with robotics and artificial intelligence is important in the EU, Spain, and the United Kingdom.

The strategies of the EU, Finland, Ireland, and Italy emphasise great potential in developing cities into new urban bioeconomy hubs. This is related to urban biowaste, the redevelopment of urban contaminated sites, and experimental smart green urban infrastructures.

Policy measures in policy strategies

Different goals, prioritisation, and policy measures can be found in bioeconomy policy strategies and their action plans across the globe. The International Advisory Council on Global Bioeconomy, in its *Global Bioeconomy Policy Report (IV)*, points out that these policy measures are often loosely defined and not supplemented with budgetary allocations. On the other hand, countries with an action plan (e.g., Spain, France, Norway, Italy, Ireland, and the EU) provide more concrete indications of what measures would be taken to support the strategy's implementation. In general, the proposed measures focus on both the demand and supply sides. In this section, a closer look is provided at the policy measures in bioeconomy strategies around the world. Generally, public investment in bioeconomy development typically includes research and innovation funding, infrastructure development, commercialisation support, bioeconomy-friendly framework conditions, demand-side instruments, capacity building and education, and measures for good governance.

Research and innovation funding

Public research and development (R&D) funding is seen as a crucial measure for innovations in the bioeconomy field. Countries support multidisciplinary research alliances and establishing research networks and centres of excellence.

They also encourage the development of bioeconomy hubs, networks, and clusters. A popular concept in Asia (Thailand, Malaysia, Japan) has emerged following "the Singaporean example of the so-called 'Biopolis', in which major research institutes come together to create a centre of excellence in bioeconomy experimentation" (International Advisory Council on Global Bioeconomy 2020b, p. 158). Austria, the EU, Germany, Italy, and the United States evolved bioeconomy monitoring approaches, while the goal of improving technology transfer remains central to all bioeconomy strategies.

Infrastructure development

Regarding infrastructure investments, there is a large-scale biorefinery development in Finland, France, Ireland, Italy, and Norway. Italy in its action plan focuses on investments in flagship projects aimed at developing infrastructures in urban biowaste and national biorefineries.

Commercialisation support

Countries in their bioeconomy strategies provide a wide spectrum of measures to support the industry. These measures range from direct financial assistance via grants, loans, cooperative agreements, tech transfer activities to subsidies and tax incentives for making bio-based products competitive. Start-up support is seen as a tool to promote bioeconomy innovation in Costa Rica, Germany, Italy, Japan, and Nordic countries (International Advisory Council on Global Bioeconomy 2020b, p. 159). Recent bioeconomy strategies focus on increasing investment and risk finance in the bioeconomy, while venture capital and investment funds for bio-based start-ups are promoted in France, the United Kingdom, Nordic countries, South Africa, and the United States.

Bioeconomy-friendly framework conditions

Bioeconomy-friendly framework conditions are frequently seen as measures to promote the bioeconomy. On one side, strategies in Ireland, Spain, the United Kingdom, and the United States focus on identifying potentially unfavourable aspects of the existing regulatory framework, barriers to introducing new products, and actions that enable rapid development and deployment of new technologies. On the other side, strategies in Costa Rica, Norway, Italy, and the United Kingdom seek to provide better incentives for bioeconomy development while focusing on policy, regulation, and industry guidance on waste. Strategies from countries and regions like Norway, Nordic countries, and Thailand focus more on environmental taxation.

Demand-side measures

Demand-side policy measures are required to stimulate change and encourage more sustainable lifestyles and consumer habits. Countries like Austria, EU,

France, Ireland, Italy, Latvia, Nordic countries, Norway, South Africa, Spain, the United Kingdom, and the United States implement public procurement policies. Furthermore, many strategies like those of Austria, the EU, Finland, France, Germany, Italy, Ireland, Japan, Latvia, Nordic countries, Norway, and Thailand promote labelling, certification, and standards for bio-based products. Although, key demand-side measure in most strategies are raising public awareness through fairs, conferences, and dialogue platforms.

Education and capacity building

The bioeconomy places special importance to knowledge, education and training, and capacity building. Proposed measures related to education and capacity building are found in almost all bioeconomy strategies. They cover programmes for lower-level school systems in Malaysia, Nordic countries, and the United States; programmes of general science, technology, engineering, and mathematics education in Italy and the United Kingdom; to bachelor, master, and apprenticeship programmes in Nordic countries and the United States. The EU, France, Italy, Ireland, Latvia, Spain, Thailand, and the United Kingdom have measures for continuing professional development while self-learning tools and open access platforms are found in the EU and Spain. Environmental education is the focus of Costa Rica's strategy, industry partnerships for early career researchers is the focus for the United Kingdom, while advanced digital skills and integrating converging technologies can be found in strategies of the EU, Germany, and Japan.

Good governance

The broad range of bioeconomy actors and their coordination represents a significant challenge to bioeconomy development. As a result, the issue of good governance has received a lot of attention in the scientific papers and at conferences and workshops (German Bioeconomy Council 2018). However, principles of good governance, namely accountability, transparency, effectiveness and coherence, participation, and fairness (Devaney *et al.* 2017), have evolved faster in national and international forums than in policy strategies (International Advisory Council on Global Bioeconomy 2020b, p. 161).

Bioeconomy policy strategies aim to increase policy coherence and effectiveness with measures of intergovernmental coordination (Germany, Costa Rica, the United States) or policy coordination (Austria and Germany). In Italy and the United Kingdom, the strategies seek to take advantage of regional expertise.

In regard to the process of strategy formulation, most countries have adopted a bottom-up, participatory approach while involving all relevant stakeholders. Regional, national, or international policy forums are considered rather significant for ensuring mutual learning and inclusive participation (International Advisory Council on Global Bioeconomy 2020b, p. 161).

To encourage interregional coordination and sharing of best practices, the updated bioeconomy EU strategy *A Sustainable Bioeconomy for Europe* planned to set up an EU bioeconomy policy support facility, with a focus on Central and Eastern Europe. Countries from Central and Eastern Europe lag behind in developing their dedicated bioeconomy strategies, despite their vast biomass potential. Therefore, 11 member states (Bulgaria, Croatia, Czech Republic, Hungary, Poland, Romania, Slovakia, Slovenia, Estonia, Lithuania, Latvia) plan to develop their bioeconomy strategy under the BIOEAST initiative and with support from the EU bioeconomy policy support facility (European Commission, SCAR, BIOEAST 2019).

Bioeconomy strategies in Ireland, Italy, and Norway promote international collaboration through strengthening international networking and cooperation between research institutions. The ones in the United States and the United Kingdom aim to improve international competitiveness. The EU and German strategies apprehend the need for regular and strategic international cooperation, and the importance of the Global Bioeconomy Summit as a platform for exchanging information between international experts beyond research (International Advisory Council on Global Bioeconomy 2020b, p. 162). The EU, Finland, France, Germany, Italy, and the Nordic countries have established bioeconomy advisory councils or stakeholder panels.

Countries with an action plan, like Ireland, Italy, Spain, Japan, and Germany, establish implementation groups that provide feedback in bioeconomy policymaking. Norway and South Africa employ collaboration committees or interdepartmental stakeholder groups. Finland has a Bioeconomy Panel, whose task is to participate in strategy formulation, review, and update, and it is also consulted on implementation. Strategies in Japan, Latvia, Malaysia, South Africa, and the United Kingdom are among the few country strategies that propose metrics and indicators for measuring success (International Advisory Council on Global Bioeconomy 2020b, p. 163).

An increasing number of regional strategies show a growing trend towards regionalisation. This is especially seen in Europe, in countries like France, Germany, Norway, Spain, and the United Kingdom, and in Latin America.

Emerging policy trends: stakeholder- and industry-driven initiatives

Besides national strategies developed by countries, stakeholder- and industry-driven strategies are also on the rise. These strategies do not represent official strategic government documents but rather aim to provide a common vision for bioeconomy development and raise awareness among political decision-makers.

In 2019, Portugal developed the *Blue BioEconomy Roadmap for Portugal*, a stakeholder roadmap endorsed by the Portuguese Ministry of Sea. This document puts Portugal at the forefront of the blue bioeconomy in Europe by 2030. In Argentina, in 2019, a collective stakeholder strategy *Bioeconomy as a*

Strategy for the Development of Argentina was published. In Canada, an industry-driven national strategy, *Canada's Bioeconomy strategy*, that represents the vision of industry representatives across the country, was developed. As a result of these industry efforts, the Canadian government invested funds in an industry-led consortium of companies, associations, and academic and research institutions from several sectors to build the national bioeconomy strategy.

Multilateral policy dialogue

Over the years, more and more multilateral bioeconomy policy initiatives have been launched under the lead of supranational and international organisations as well as multi-stakeholder initiatives. This section brings an overview of a few main attempts that have been made to establish structures for multilateral policy dialogue with the aim of fostering the development of a sustainable bioeconomy: the International Knowledge-Based Bioeconomy Forum, the European Commission's International Bioeconomy Forum, the UN Food and Agriculture Organization (FAO) International Sustainable Bioeconomy Working Group and the BioFuture Platform.

The International Knowledge-Based Bioeconomy Forum is a scientific cooperation initiative between the European Commission, Australia, Canada, and New Zealand. It was launched in 2010, and its main goal is to improve "the research and innovation policy dialogue and scientific cooperation between the four partners regarding the most important issues of the bioeconomy" (European Commission 2022).

The International Bioeconomy Forum is a mechanism for a long-term R&D collaboration in the bioeconomy, initiated by the European Commission in 2016. The International Bioeconomy Forum focuses on building policy coherence and aims at exploiting synergies among countries, regions, and sectors. The forum is organised in ad hoc working groups, which cover four areas: Food Systems Microbiome, Information and Communication Technology in Precision Food Systems, Plant Health, and the Forest Bioeconomy (International Bioeconomy Forum 2021).

The International Sustainable Bioeconomy Working Group is a UN FAO-led platform established in 2015 to support countries in developing sustainable and circular bioeconomy strategies, action plans, and programmes in line with the SDGs and the Paris Agreement. It presents an international, multi-stakeholder expert group from all five continents. The Working Group covers three main areas: share lessons learnt and good practices on the potential benefits and risks of food system bioinnovations, provide guidance to national and regional stakeholders in the development of sustainable and circular bioeconomy strategies, and support bioeconomy monitoring and evaluation (Food and Agriculture Organization of the United Nations (UN FAO) 2022). The International Sustainable Bioeconomy Working Group is considered an important South-South and Triangular Cooperation platform within the UN FAO that supports the dissemination of sustainable bioeconomy in developing countries.

The BioFuture Platform is a government-led, multi-stakeholder initiative that was launched in 2016, during COP22 in Marrakesh. In 2019, the International Energy Agency assumed the role of the secretariat and organised ministerial-level meetings and summits to promote more consistent international collaboration and dialogue.

Macro-regional actors and policy initiatives

Along with national and international actors, there are a number of regions and macro-regions that are active in the bioeconomy policy field. This section provides an overview of some macro-regional actors and policy initiatives around the world.

In Africa, East African Regional Strategy was published in 2020. It was a result of the BiSEA project that aimed to develop a regional, innovation-driven bioeconomy strategy. With the help of the Africa Centre of the Stockholm Environment Institute and in close consultations with partners from African countries (Ethiopia, Burundi, Kenya, Rwanda, Tanzania, Uganda, and South Sudan), the first African regional strategy was successfully finalized.

In Europe, the Central and Eastern European Bioregions Forum was founded in 2016. It represented a continuation of the European Bioeconomy Congress (EBCL) held in Lodz, where the strategic document for bioeconomy development in local Central and Eastern Europe *Lodz Declaration of Bioregions* (EBCL 2016) was signed. The document promotes the concept of development of bioeconomy in local bio-communities (bio-villages, bio-cities, and bioregions).

Another initiative in Eastern Europe is the Central-Eastern European Initiative for Knowledge-based Agriculture, Aquaculture and Forestry in the Bioeconomy (BIOEAST 2018). Activities for this initiative were started by Visegrád Group countries (Czech Republic, Hungary, Poland, and Slovakia) and were then joined by Bulgaria, Croatia, Latvia, Lithuania, Estonia, Romania, and Slovenia. Today it aims to work towards sustainable bioeconomies in all countries of Central and Eastern Europe (ed. Keswani 2020).

The Bioeconomy Strategic Working Group was launched as a thematic working group under the Standing Committee on Agricultural Research which advises the European Commission and the member states on agricultural research in Europe. The Bioeconomy Strategic Working Group discuses technical and strategical issues related to the bioeconomy and aims to facilitate informal exchanges between the European member states. (BIOEAST 2020).

The European Bioeconomy Policy Forum was established in 2020 as a new dedicated forum for member states to support the strategic advancement of a circular and sustainable bioeconomy in Europe. It brings together member states and EU institutions to facilitate sharing best practices and coordinated approaches in developing bioeconomy policy solutions. It was established under Action 2.3 of the updated European Bioeconomy Strategy in response to member states' requests for increased cooperation on bioeconomy policy (European Commission 2020b).

In Northern Europe, the Nordic Council of Ministers, founded in 2012, involves countries like Denmark, Finland, Iceland, Norway, Sweden, the Faroe Islands, Greenland, and Åland. It promotes bioeconomy development as an official body for intergovernmental cooperation in the Nordic region (Mathisen 2019).

In Latin America, the UN Economic Commission for Latin America and the Caribbean is a driving force of the macro-regional bioeconomy development. It organises events and conducts technical assistance missions in countries like Costa Rica, Uruguay, and Argentina with a goal of designing the national bioeconomy strategies. In 2019 the Latin American Bioeconomy Network was founded to advance the exchange of experiences and to promote the bioeconomy policies.

Conclusions

The trend of developing bioeconomy policy strategies is on the rise, and today there are almost 60 countries across the globe that already developed their bioeconomy-related policies. Nineteen dedicated national and macro-regional bioeconomy strategies have been published so far. Countries pursue their goals and adopt policy documents adapted to local conditions.

The expectations associated with the bioeconomy have changed considerably in recent years and, in many respects, have increased. Since the adoption of the 2030 Agenda, the relationship between the SDGs and the bioeconomy has strengthened. Additionally, recent policy documents look at the bioeconomy as a new strategy for reindustrialisation and coping with global political challenges, such as climate change mitigation, the Covid-19 pandemic, and biodiversity issues.

When countries adopt their bioeconomy strategies, they lay out the foundation for policy support and investments. These investments enable pioneering research and the development of new technologies. They push forward industrialisation processes and help spike consumer demand. Moreover, they support education and capacity building and create awareness about the importance of the bioeconomy.

More recent strategies consider a wider range of objectives, and there is no longer a clear hierarchy but rather a diverse set of equal goals. As a consequence of this growing heterogeneity of goals, more recently updated strategies search for amplified impact by better aligning with other policy priorities. Nevertheless, many of these goals have not been translated into specifically implemented policy measures. Only half of the countries analysed have published dedicated action plans to accompany their bioeconomy strategies, and those action plans vary to a large extent. They spread from action plans with concrete policy measures with indicators and milestones to more general progress reports.

There is a trend towards increasing thematic prioritisation and specialisation in the policy documents. Different countries concentrate on different priorities and specialise in different sectors. This largely depends on the countries' natural resources, technological capacity and comparative economic advantage. Accordingly, the bioeconomy continues to be seen as diverse, showing that there is not one bioeconomy but many.

There is a wide range of bioeconomy actors across the globe. Coordinating their activity represents a sizable challenge to bioeconomy development. Consequently, multilateral structures and principles of good governance have become more relevant than ever before. Also, an increasing number of multi-stakeholder initiatives are seen as furthering bioeconomy development. In their activities they encourage country's specific vision of the bioeconomy. In addition, there is an expanding involvement of international and macro-regional actors and regional activities that drive the bioeconomy development.

Bibliography

Albrecht, K., and Ettling, S., 2014. Bioeconomy strategies across the globe. *Rural 21*, 48 (3), 10–13. Available from: www.rural21.com/fileadmin/downloads/2014/en-03/rural2014_03-S10-13.pdf [Accessed 17 December 2021].

BIOEAST, 2018. Central and Eastern European initiative for knowledge-based agriculture, aquaculture and forestry in the bioeconomy. Available from: https://bioeast.eu/home/ [Accessed 16 December 2021].

BIOEAST, 2020. Bioeconomy and Green Investments: Unique opportunity to tackle the crisis in the Central and Eastern European countries. Available from: https://bioeast.eu/wp-content/uploads/2020/06/BIOEAST-Position-2020-06-10.pdf [Accessed 16 December 2021].

Bogdanski, A., Giuntoli, J., Mubareka, S., Gomez San Juan, M., Robert, N., and Tani, A., 2021. *Guidance Note on Monitoring the Sustainability of the Bioeconomy at a Country or Macro-regional Level* (Environment and Natural Resources Management Working Papers – Bioeconomy, no. 90). Rome: FAO and EC-JRC. Available from: https://doi.org/10.4060/cb7437en [Accessed 15 December 2021].

Council for Integrated Innovation Strategy, 2020. Bio-strategy 2020. Available from: https://www8. cao.go.jp/cstp/bio/bio2020_honbun.pdf [Accessed 28 December 2021].

Devaney, L., Henchion, M., and Regan, Á., 2017. Governance in the bioeconomy. *EuroChoices*, 6 (2), 41–46. Available from: https://doi.org/10.1111/1746-692X.12141

EBCL, 2016. Lodz declaration of bioregions. Available from: www.sureaqua.no/Sureaqua/library/Lodz%20Declaration%20of%20Bioregions,%202016.pdf [Accessed 16 December 2021].

Economic Commission for Latin America (ECLA), n.d. Bioeconomy. Available from: www.cepal.org/en/topics/bioeconomy [Accessed 20 December 2021].

European Commission, 2012. Innovating for sustainable growth: A bioeconomy for Europe. Available from: https://op.europa.eu/en/publication-detail/-/publication/1f0d8515-8dc0-4435-ba53–9570e47dbd51 [Accessed 10 December 2021].

European Commission, 2018a. A sustainable bioeconomy forr Europe: Strengthening the connection between economy, society and the environment: Updated bioeconomy strategy. Available from: https://data.europa.eu/doi/10.2777/478385 [Accessed 15 December 2021].

European Commission, 2018b. Bioeconomy: The European way to use our natural resources, Action plan 2018. Available from: https://europanels.org/wp-content/uploads/2021/02/COM-Bioeconomy-Action-Plan-2018.pdf [Accessed 28 December 2021].

European Commission, 2019. The European green deal. Available from: https://ec.europa.eu/info/sites/info/files/european-green-deal-communication_en.pdf [Accessed 15 December 2021].

European Commission, 2020a. A new circular economy action plan for a cleaner and more competitive Europe. Available from: https://eur-lex.europa.eu/legal-content/EN/TXT/HTML/?uri=CELEX:52020DC0098&qid=1608196909801&from=EN [Accessed 15 December 2021].

European Commission, 2020b. High-level launch of the European Bioeconomy Policy Forum. Available from: https://ec.europa.eu/info/events/high-level-launch-european-bioeconomy-policy-forum-2020-nov-12_en [Accessed 16 December 2021].

European Commission, 2022. International knowledge-based bioeconomy forum. Available from: https://ec.europa.eu/research/bioeconomy/international-cooperation/forum/index_en.htm [Accessed 6 January 2022].

European Commission, Joint Research Centre, 2018. Brief on jobs and growth of the bioeconomy 2009–2015. Available from: https://data.europa.eu/doi/10.2760/718006 [Accessed 17 December 2021].

European Commission, SCAR – Bioeconomy Strategic Working Group, BIOEAST Initiative, 2019. Joint-workshop Policy Support Facility (PSF) on bioeconomy. Available from: www.scar-swgsbgb.eu/lw_resource/datapool/_items/item_37/first_psf_workshop_13_03_2019_final_report.pdf [Accessed 6 January 2022].

European Commission's Knowledge Centre for Bioeconomy, 2022. Bioeconomy country dashboard. Available from: https://knowledge4policy.ec.europa.eu/visualisation/bioeconomy-different-countries_en [Accessed 11 January 2022].

European Environmental Agency (EEA), 2019. Sustainability transitions: Policy and practice. Available from: www.eea.europa.eu/publications/ sustainability-transitions-policy-and-practice [Accessed 26 November 2021].

Flemish Government, Department of Economy, Science and Innovation, 2020. Flanders Bioeconomy – An unfolding story of sustainable growth. Available from: www.ewivlaanderen.be/sites/default/files/bestanden/flanders_bioeconomy_2020.pdf [Accessed 15 December 2021].

Food and Agriculture Organization of the United Nations (UN FAO), 2016. FAO's south-south and triangular cooperation strategy in action. Available from: www.fao.org/3/a-i6249e.pdf [Accessed 15 December 2021].

Food and Agriculture Organization of the United Nations (UN FAO), 2022. Sustainable and circular bioeconomy for food systems transformation. Available from: www.fao.org/in-action/sustainable-and-circular-bioeconomy/international-sustainable-bioeconomy-working-group/en/ [Accessed 6 January 2022].

German Bioeconomy Council, 2015a. Bioeconomy policy. Synopsis and analysis of strategies in the G7. Available from: https://gbs2020.net/wp-content/uploads/2020/04/BOER_Laenderstudie_1_.pdf [Accessed 12 November 2021].

German Bioeconomy Council, 2015b. Bioeconomy policy (Part II). Synopsis of national strategies around the world. Available from: https://knowledge4policy.ec.europa.eu/publication/bioeconomy-policy-part-ii-synopsis-national-strategies-around-world_en [Accessed 12 November 2021].

German Bioeconomy Council, 2018. Bioeconomy Policy (Part III) update report of national strategies around the world. Available from: www.biooekonomierat.de/media/pdf/archiv/international-bioeconomy-policy-part-III.pdf?m=1637834907& [Accessed 12 November 2021].

German Federal Ministry of Education and Research, 2011. National research strategy bioeconomy 2030. Available from: http:// biotech2030.ru/wp-content/uploads/docs/int/bioeconomy_2030_germany. pdf [Accessed 15 November 2021].

German Federal Ministry of Food and Agriculture, 2014. National policy strategy on bioeconomy. Available from: www.bioways.eu/download.php?f=62&l=en&key=c21c2ea7e095424f3545c66da7b98821 [Accessed 15 November 2021].

Heimann, T., 2019. Bioeconomy and SDGs: Does the bioeconomy support the achievement of the SDGs? *Earth's Future*, 7, 43–57. Available from: https://doi.org/10.1029/2018ef001014 [Accessed 17 December 2021].

International Advisory Council on Global Bioeconomy, 2020a. Expanding the sustainable bioeconomy – Vision and way forward. Communiqué of the global bioeconomy summit 2020. Available from: https://knowledge4policy.ec.europa.eu/publication/expanding-sustainable-bioeconomy-%E2%80%93-vision-way-forward-communiqu%C3%A9-global-bioeconomy_en [Accessed 12 November 2021].

International Advisory Council on Global Bioeconomy, 2020b. Global Bioeconomy Policy Report (IV): A decade of bioeconomy policy development around the world. Available from: https://gbs2020.net/wp-content/uploads/2020/11/GBS-2020_Global-Bioeconomy-Policy-Report_IV_web.pdf [Accessed 12 November 2021].

International Bioeconomy Forum, 2021. International bioeconomy forum. Available from: www.bioeconomy-forum.org/ [Accessed 16 December 2021].

Keswani, C., ed., 2020. *Bioeconomy for Sustainable Development*. Singapore: Springer Nature. Available from: https://doi.org/10.1007/978-981-13-9431-7 [Accessed 17 December 2021].

Lange, L., Connor, K. O., Arason, S., Bundgård-Jørgensen, U., Canalis, A., Carrez, D., Gallagher, J., Gøtke, N., Huyghe, C., Jarry, B., Llorente, P., Marinova, M., Martins, L. O., Mengal, P., Paiano, P., Panoutsou, C., Rodrigues, L., Stengel, D. B., van der Meer, Y., and Vieira, H., 2021. Developing a sustainable and circular bio-based economy in EU: By partnering across sectors, upscaling and using new knowledge faster, and for the benefit of climate, environment & biodiversity, and people & business. *Frontiers in Bioengineering and Biotechnology*. Available from: https://doi.org/10.3389/fbioe.2020.619066 [Accessed 13 December 2021].

Mathisen, M., 2019. Nordic bioeconomy programme. Available from: https://colloque.inrae.fr/bioeconomy2019/content/download/4055/44494/version/1/file/04.+M-MATHISEN+nordic_bioeconomy_strategy.pdf [Accessed 20 December 2021].

M'barek, R., Philippidis, G., and Ronzon, T., 2019. Alternative global transition pathways to 2050: Prospects for the bioeconomy – An application of the MAGNET model with SDG insights. Available from: https://doi.org/10.2760/594847 [Accessed 10 December 2021].

National Bioeconomy Coordination Board of the Presidency of Council of Ministers, 2021. Implementation Action Plan (2020–2025) for the Italian Bioeconomy Strategy BIT II. Available from: https://cnbbsv.palazzochigi.it/media/2079/iap_2332021.pdf [Accessed 28 December 2021].

Nordic Council of Ministers, 2018. Nordic bioeconomy programme: 15 action points for sustainable change. Available from: http://norden.diva-portal.org/smash/get/diva2:1222743/FULLTEXT01.pdf [Accessed 13 December 2021].

Organisation for Economic Cooperation and Development (OECD), 2009. The bioeconomy to 2030: Designing a policy agenda. Available from: https://doi.org/10.1787/9789264056886-en [Accessed 28 December 2021].

Organisation for Economic Cooperation and Development (OECD), 2018. Meeting policy challenges for a sustainable bioeconomy. Available from: https://doi.org/10.1787/9789264292345-en [Accessed 12 December 2021].

Patermanna, C., and Aguilar, A., 2018. The origins of the bioeconomy in the European Union. *New Biotechnology*, 40 (part A), 20–24. Available from: https://doi.org/10.1016/j.nbt.2017.04.002 [Accessed 12 December 2021].

Philippidis, G., Bartelings, H., Helming, J., M'barek, R., Ronzon, T., Smeets, E., Van Meijl, H. and Shutes, L., 2018. The MAGNET Model framework for assessing policy coherence and SDGs: Application to the bioeconomy. Available from: https://publications.jrc.ec.europa.eu/repository/handle/JRC111508 [Accessed 17 December 2021].

Philp, J., 2018. The bioeconomy, the challenge of the century for policy makers. *New Biotechnology*, 40 (part A), 11–19. Available from: https://doi.org/10.1016/j.nbt.2017.04.004 [Accessed 12 December 2021].

Pietzsch, J., and Schurr, U., 2017. Introduction. In J. Pietzsch, ed. *Bioeconomy for Beginners*. Berlin: Springer-Verlag, 1–10.

Ronzon, T., Iost, S., and Philippidis, G., 2022. An output-based measurement of EU bioeconomy services: Marrying statistics with policy insight. *Structural Change and Economic Dynamics*, 60, 290–301. Available from: https://doi.org/10.1016/j.strueco.2021.10.005 [Accessed 11 December 2021].

Schütte, G., 2018. What kind of innovation policy does the bioeconomy need? *New Biotechnology*, 40 (part A), 82–86. Available from: https://doi.org/10.1016/j.nbt.2017.04.003 [Accessed 10 November 2021].

Staffas, L., Gustavsson, M., and McCormick, K., 2013. Strategies and policies for the bioeconomy and bio-based economy: An analysis of official national approaches. *Sustainability*, 5 (6), 2751–2769. Available from: https://doi.org/10.3390/su5062751 [Accessed 10 December 2021].

The White House, 2012. National bioeconomy blueprint. Available from: https://obamawhitehouse.archives.gov/sites/default/files/microsites/ostp/national_bioeconomy_blueprint_april_2012.pdf [Accessed 15 December 2021].

UN, 2015. Transforming our world, the 2030 Agenda for sustainable development. Available from: https://sdgs.un.org/sites/default/files/publications/21252030%20Agenda%20for%20Sustainable%20Development%20web.pdf [Accessed 16 December 2021].

UN, 2016. Paris Agreement. Available from: https://unfccc.int/process-and-meetings/the-paris-agreement/the-paris-agreement [Accessed 16 December 2021].

United Nations Office for South-South Cooperation, 2019. South-south and triangular cooperation on the bioeconomy in light of the Paris Agreement and the 2030 Agenda for sustainable development. Available from: https://www.unsouthsouth.org/wp-content/uploads/2019/12/Bioeconomy-Publication_visualization-for-website.pdf [Accessed 15 December 2021].

3 Green and blue economy

Anamarija Pisarović

Introduction

Since the concept of the linear economy becomes unsustainable due to the premise that resources are inexhaustible and space for waste disposal is unlimited, the circular economy can be defined as a transition from a current linear economy with existing resources to an economy where the raw materials and products remain in a closed cycle by combining all production chains and not just one (D'Amato *et al.* 2017). The concept of a bioeconomy is linked to the idea of a circular economy, which is based on the reuse and recycling of various products (not just bioproducts) and includes all types of material streams that enter the cycle to be reused (Carus 2017). The main goal of the bioeconomy is to use existing biomass sustainably and economically while keeping carbon emissions to a minimum by replacing fossil carbon with renewable carbon. At the same time, the circular economy focuses on reusing products by recycling and closing the loop within large cycles (European Environment Agency 2016). The bioeconomy focuses on connecting the environment, society, and economy and covers all sectors and systems that rely on biological resources by connecting terrestrial, aquatic, and marine ecosystems, all primary production sectors that use and produce biological resources, such as agriculture, forestry, fisheries, and aquaculture, and all economic and industrial sectors that use biological resources for production, energy, and services. A thriving bioeconomy implies sustainability along with circular and cascading use of resources with a focus on local specifics to drive industrial renewal, modernise primary production systems, protect the environment, and preserve biodiversity (El-Chichakli *et al.* 2016). Because it is based primarily on biogenic and not fossil resources, the bioeconomy represents a substantial innovation potential in the field of science and industrial technologies. The European Union (EU) in 2015 adopted a circular economy strategy to promote resource efficiency in all industries and member states (European Commission 2015), thus making the bioeconomy, as part of the circular economy, an increasingly prevalent economic sector in policymaking worldwide.

Environmental policies affect the value chain in such a way that the established standards encourage activities that will positively impact the development of "green" industries while limiting activities harmful to the environment.

DOI: 10.4324/9781003223733-4

Transport policies can threaten the global competitiveness of the bioeconomy due to logistical costs while paving the way for a circular economy based on national supply chains. As the bioeconomy offers solutions to comprehensive societal challenges, such as food security, climate change, limited natural resources (including fossil fuels), economic growth, and demand for raw materials, most Organisation for Economic Development countries already have national bioeconomy strategies in place to support the United Nations (UN) Sustainable Development Goals (SDGs) (United Nations 2015a) and obligations under the Paris Climate Agreement (United Nations 2015b). Green and blue economies aim at preserving natural capital and reducing environmental risks detrimental to terrestrial, marine, and aquatic ecosystems with a view to sustainable development and prevention of environmental degradation (D'Amato *et al.* 2017). The common goals of these concepts (bioeconomy, circular economy, green and blue economy) are in line with climate and energy goals. These entail a more sustainable and efficient world with a low carbon footprint, as well as avoiding the use of additional fossil carbon.

Forests are the most diverse ecosystem on Earth, offering unique and varied habitats to plants, animals, fungi, and microorganisms. Forests are home to more than 80% of all terrestrial species globally and are crucial for preserving biodiversity. One of the leading causes of global warming is excess CO_2 of anthropogenic origin in the atmosphere. Indeed, as they store carbon throughout their entire life cycle, forests serve as a carbon sink. Once the trees die and decompose naturally, they release carbon back into the atmosphere and such a carbon cycle is considered neutral. Furthermore, if wood from sustainably managed forests is used as a raw material for construction, it continues to retain carbon, and the products are considered environmentally friendly. Today, forests and the forest-based sector play a central role in the European bioeconomy as they provide material (wood and non-wood products), bioenergy, and a wide range of other ecosystem services (European Commission 2021).

Agriculture provides healthy and safe food, feed, raw materials for liquid fuels, and value-added products. As such, it has a solid potential to occupy a central place in the bioeconomy. On the other hand, the bioeconomy has a number of potential applications in agriculture in terms of cultivating dedicated energy crops due to high energy yields, crops that consume less water or can be irrigated with wastewater, by efficiently using fertilisers and planting crops more resistant to diseases that increase yields and produce less CO_2. The cultivation of dedicated energy crops is expected to expand due to the high energy yield, which is approximately three times higher than the yield of traditional crops. These crops can be irrigated with wastewater, thus reducing the pressure on the environment.

As the world's largest ecosystem and the primary climate regulator at our disposal, the ocean represents the "lungs of the planet", provides food for about half of the world's population, and represents the habitat for about 80% of the Earth's biodiversity. Most industries like shipping, fishing, aquaculture, and coastal tourism depend on ocean health. The blue economy is the cornerstone of both the European Green Deal and the Recovery plan for Europe, which will define the European economy in the future. The blue economy should adhere, like any other sector, to the European Green Agreement. "Blue is a prerequisite for the green" given that the blue economy contributes to climate change mitigation by developing offshore renewable energy sources, decarbonising maritime transport and greening ports, and that developing green infrastructure in coastal areas will help preserve biodiversity and landscapes, thus benefitting tourism and the coastal economy. Ship recycling and renewing fishing equipment design standards will contribute to the circular economy.

Forest-based bioeconomy

A forest-based bioeconomy is characterised by sustainable forest management focused on the use of wood and non-wood forest products as well as forest services such as recreation, tourism, and clean environment. It comprises those parts of the economy that sustainably produce renewable biological resources and convert them into value-added products and services, thereby reducing dependence on fossil fuels.

EU member states have responsibly approached the conservation and sustainable management of forests as an important environmental challenge of our time, adopting the forestry strategy for the EU proposed by the European Commission in 1998 (European Commission 1998). In response to the additional challenges forests and the forestry sector faced, the New EU Forest Strategy (European Commission 2013) was adopted in 2013. The ecological importance of forests for the carbon cycle as well as providing habitats to various plant and animal species and regulating the water regime is crucial for survival. Due to the strong European influence on the international stage, EU member states have become role models of forest protection and sustainable forest management globally. The successful establishment of the Natura 2000 network (European Commission 2003) and other initiatives such as national forestry programmes and the implementation of the resolutions of the Ministerial Conference on Forest Protection in Europe represent important achievements for the EU (European Commission n.d.). As a pan-European high-level voluntary policy process for intergovernmental dialogue and cooperation on forest policies in Europe, Forest Europe brings together 46 member states of the Council of Europe. It jointly develops pan-European criteria and indicators for sustainable forest management as instruments to help protect and sustainably manage forests (Wolfslehner *et al.* 2016). The criteria include maintaining and enhancing forest resources and their contribution to the global carbon cycles, maintaining the health and vitality of the forest ecosystem, maintaining and enhancing the productive functions of forests (wood and non-wood), maintaining and conserving the biodiversity of forest ecosystems, maintaining and enhancing the protective functions in forest management (especially soil and water), as well as maintaining other socio-economic functions and conditions (Forest Europe 2015).

Forests and forestry in the EU are characterised by a wide range of climatic, geographical, ecological, and socio-economic conditions. Approximately 70% of the forest area is located in Finland, France, Germany, and Sweden. Nevertheless, Southern Europe has the most significant potential for conserving and restoring forest biodiversity. The Mediterranean biogeographical area hosts an incredible number of 30,000 vascular plants, of which over 10,000 are exclusively regional, making it one of the richest areas in terms of endemic species. Sweden, Finland, Germany, France, and Austria are among the world's ten largest exporters of forest products. Nevertheless, many European forests are underutilised, so a general phenomenon of forest biomass growth has been observed as annual growth exceeds felling. Therefore, very dense forest stands predominate in many EU countries. Forest ownership varies within the EU. In Greece and Ireland, the state owns about two-thirds of the forest land, while in Belgium, Spain, Italy, Luxembourg, France, and Germany, local communities play an essential role as forest owners. Forest cover in the EU is increasing, both through EU afforestation programmes and from natural succession on abandoned land that has been used for grazing in the past. Given the awareness of the importance of tropical forests and the harmfulness of deforestation, since 2000, the demand for tropical wood products has decreased. As a net exporter of wood products, Europe has become one of the world's leading producers of log wood. Nearly 3 million people in Europe earn a living working in forestry and forest-based industries. The new EU forest strategy for 2030 takes the lead in the implementation of the European Green Deal and the EU biodiversity strategy for 2030 (European Commission 2021a). Forests contain three-quarters of the Earth's biomass, half of which is carbon, and therefore play a vital role in the global carbon cycle, directly affecting the climate. The value chain begins with carbon storage as a critical ecosystem service that captures atmospheric carbon into other carbon sinks such as forest vegetation and soil. Through photosynthesis, trees absorb carbon dioxide and release oxygen while using carbon to build woody parts, trees, branches, roots, and leaves. On the other hand, trees release carbon dioxide through respiration and through decomposition or combustion processes after they die. It is estimated that for each tonne of wood used instead of non-wood products, there is an average emission reduction of 2 tonnes of carbon. Moreover, wood production stores carbon throughout its lifetime, so the positive impact on the environment is long-lasting. At the end of its life, a wood product can be recycled and reused as a new product or used for energy (Ontl and Schulte 2012).

A forest-based value chain connects different processes, such as processing, design, production, and sale of wood products and their components. In addition to combining several production processes that transform natural forest resources into products and services, it also connects the use of wood residue and wood waste for energy generation, opens numerous businesses, and raises awareness of the importance of wood as a material or energy source. Most wood is obtained from forests by regular or sanitary felling and plantations of fast-growing species of trees and shrubs. Forests offer many benefits, from drinking water and biodiversity to recreation and tourism, that are of great importance for the common good and the economy. However, some cannot be easily valued (hydrological function, aesthetic and landscape services, etc.). In forest management, it is essential to consider the diversity of forest functions, so forest management, where the role of forests in the ecosystem is emphasised, is different from that which focuses on timber production. Some forest functions (such as log production, hunting, non-timber forest products) are more easily valued in economic terms than others. In addition to tangible forest products (wood and non-wood), forests provide various services such as socio-cultural, educational, aesthetic and landscape services, as well as ecological forest functions. They are essential for social well-being and spiritual benefits (Forest Europe 2019). The ecological functions of forests are particularly difficult to evaluate. Still, these roles are particularly important because they ensure a healthy environment, ensure clean water, and help fight global warming, so they are crucial to our well-being. In addition to biodiversity loss, climate change and global warming are some of today's biggest challenges (Millennium Ecosystem Assessment 2005; Mapping and Assessment of Ecosystems and Their Services 2013). By regulating surface and underwater flows, forests help preserve water supplies and prevent soil erosion, especially in mountainous areas and climate extremes, thus regulating the hydrological cycle. Wood is historically one of the most important building materials due to its properties such as durability and usability. Even today, it represents the first choice for construction in many countries and is more environmentally friendly than other materials. Wood has always been an integral part of human society, and visions of a sustainable future include wood because of its important role as a renewable energy source. A forest-based bioeconomy has strong innovation potential, and research and development (R&D) of new ways of using wood are more diverse than ever. Innovation plays an essential role in the complex transition to a sustainable bioeconomy, both on the side of innovative production technologies that use bioresources as raw materials and on the side of innovative consumption options, resulting in close cooperation between stakeholders from various sectors: companies, research institutions, government bodies, associations, entrepreneurs, and users sharing knowledge, experience, and vision. Recycling and recovering waste wood into a usable product reduces the need for logging, thereby reducing pressure on forests, energy loss, and soil emissions as rotting wood produces methane harmful to the environment. In addition, recovery can be a driver of social innovation.

Agri-based bioeconomy

The agri-food sector plays a central role in most bioeconomy strategies due to agriculture's dependence on biological resources, as agriculture represents a significant source of biomass for the food industry and other bio-based industries. Accordingly, sustainable production and exploitation of biological resources contribute to food security, sustainable management of natural resources, reduction of dependency on non-renewable resources, climate change mitigation and adaptation, and job creation and competitiveness (Organisation for Economic Co-operation and Development 2018, 2019; Food and Agriculture Organization of the United Nations 2018; Staffas *et al.* 2013; Priefer *et al.* 2017).

The bioeconomy is considered key to achieving the UN SDGs addressing food security and improved nutrition, health and well-being, and clean water and sanitation, as sustainable production and exploitation of biological resources contribute to food security and sustainable management of natural resources (Von Braun 2018).

The bioeconomy provides several opportunities for the agricultural sector, the most important of which are new income and job flows by using new resources and opening new markets; diversification of agricultural practices and establishing links with new sectors and enterprises; reducing risk exposure due to policy changes by moving to more efficient, more resource-efficient business models; and reducing costs through more efficient use of resources and optimisation of waste use. New technologies could support better productivity, efficiency, and resilience while reducing environmental impact. New plant varieties, together with improved management practices, would allow crops to be better adapted to the conditions of cultivation and improvement of yields while reducing soil erosion and salinity as well as irrigation needs. As far as climate change is concerned, national strategies see an excellent opportunity to reduce greenhouse gases (GHGs) generated from agriculture. New energy sources would replace fossil fuel–based products, and the sustainable intensification of agriculture and forestry would help protect carbon stocks in soils and forests (Diakosavvas and Freza 2019).

Sustainable agri-food sectors must face the challenge of producing increasing amounts of biomass while reducing the negative impact on the environment. Agriculture and food, beverage, and tobacco production are dominant sectors of the European bioeconomy, so EU bioeconomy strategies aim to place a strong emphasis on these two sectors to provide knowledge and tools for productive, resource-efficient, and resilient bio-based food, feed, and raw materials systems, together with policies supporting life in rural areas. Increased resource efficiency and waste reduction would boost the food industry's competitiveness by reducing costs while positively impacting the environment. The development of biodegradable food packaging that can be fully reused, recycled, or recovered reduces the food industry's environmental footprint, contributes to enhanced food safety, and increases the competitiveness of the food and packaging industry (Staffas *et al.* 2013). The main idea is to make full use of all biomass produced, which can be achieved by adopting an integrated approach to crop and livestock production in a circular production system, which maximises overall food production while reducing environmental impact and improving soil carbon sequestration. This mainly entails adding value to multipurpose crops and animal manure. Adding value to a multipurpose crop means making the most of crop biomass by increasing synergies between food and feed and food and non-food products, and entails producing biomass for food, feed, and non-food purposes from the same crop. Smart food processing can allow the extraction of all biomass that can be converted into food. The remaining biomass (which animals can digest but humans cannot) can be converted into feed. The resulting waste with nutritional value can be composted and used to cultivate insects and/or fungi. Insects, in turn, can be used as a source of protein for food and feed. Adding value to animal manure is achieved by "loading" the soil with organic matter to increase carbon sequestration. Circular systems in livestock and farming would lead to reducing GHG emissions due to lower intensity of product emissions resulting from the same culture producing more products due to increased carbon sequestration and better management of livestock production. Reducing waste in the agri-food chain will require increased cross-sectoral integration and better communication and coordination in the value chain. Bioeconomy strategies point out that biowaste, including agri-food waste, is currently underused and underestimated, with possible negative consequences for the environment. By adopting innovative processes, by-products and waste from the agri-food industry can be used and marketed as new food or feed or transformed into biodegradable food packaging. Agri-food waste is extremely important for composting as it reduces the depletion of organic matter in the soil when used as fertiliser. Strategies, therefore, focus on promoting R&D of new and improved bio-based products. Many countries consider developing biorefineries essential for transforming bio-based resources into innovative products. It is equally important to develop new alternative resources for industrial use, such as organic waste and residues as well as by-products from agriculture, forestry, and fisheries.

The EU strategy emphasises the need to raise consumer awareness of the link between food and health and encourage greater information about food choices. Innovations in the agri-food sector should contribute to better and more nutritionally adapted diet and healthcare, as tailor-made and targeted diets will have substantial market potential in addressing obesity and ageing populations (European Commission 2019). In this context, the importance of R&D of functional foods (potentially with a specific positive effect on health) is highlighted, emphasising healthy diets. With the development of the bioeconomy, the food industry could take advantage of alternative food sources, such as insects and algae, which can be used as protein sources in the food and feed industry. The need to develop fast network tools to control pathogens, allergens, toxins, chemical residues, nanomaterials, and so forth, to increase food and feed safety was emphasised, so greater integration and communication throughout the supply chain are considered necessary. To improve the uptake of innovation, measures have been taken to transfer knowledge and technology to farmers, bridging the gap between research and practice and making it easier for farmers to adapt and adopt new technologies. At the same time, it is vital to recognise the potential farmers and small and medium-sized enterprises (SME) have in developing and disseminating bio-innovation by emphasising the importance of knowledge exchange instead of knowledge transfer through involving a broad range of stakeholders (farmers, foresters, fishermen, advisory services, all industries, consumers) in the supply chain as recommended by the Standing Committee on Agricultural Research-Agricultural Knowledge and Innovation System (SCAR-AKIS) working group on knowledge and innovation systems in agriculture (Standing Committee on Agricultural Research-Agriculture Knowledge and Innovation System, SCAR-AKIS 2017).

Blue bioeconomy

The blue bioeconomy is focused on producing food, feed, nutraceuticals, pharmaceuticals, cosmetics, energy, packaging, clothing, and so on, from aquatic biomass. Blue biotechnology is the application of science and technology to living aquatic organisms to produce knowledge, goods, and services. The blue bioecoomy is developing rapidly in Europe thanks to the emphasis on research and the strong engagement of all stakeholders. However, its development is uneven since innovative products are only available on the market in some regions and countries. To ensure a level playing field, the European Commission is working with European stakeholders in the framework of the Blue Bioeconomy Forum. Furthermore, a comprehensive EU algae initiative, planned for 2022, will boost the blue bioeconomy sector by supporting research and innovation, supporting data collecting and mapping, improving the regulatory and governance framework, and supporting market development (European Commission 2021b). Certain groups of marine organisms are becoming commercially interesting for the growing sectors of the blue bioeconomy and biotechnology. These include microorganisms (microalgae, bacteria, and fungi), algae, and invertebrates (e.g., starfish, sea cucumbers, sea urchins). Many activities, such as fishing, aquaculture, shipping, the use of renewable energy sources, and nature protection, compete for maritime space. To regulate these activities, the EU has adopted a directive on Maritime Spatial Planning (European Union 2014) to coherently manage the use of seas and oceans and ensure human activities take place in an efficient, safe, and sustainable way. The directive aims at reducing potential conflicts and creating synergies between different activities; encouraging investment through predictability, transparency, and legal certainty; strengthening cross-border cooperation between the EU member states to develop renewable energy sources, allocate waterways, and lay pipelines and submarine cables; and protecting the environment by assigning protected areas, calculating environmental impacts, and identifying opportunities for multiple uses of space. As maritime space must be managed in a sustainable way around the globe, the EU is working with the Intergovernmental Oceanographic Commission of UNESCO to accelerate the Maritime Spatial Planning processes worldwide through the MSPGlobal project and by developing international guidelines.

The aquaculture sector is one of the fastest-growing sectors globally. Europe is one of the largest markets for seafood products and the second-largest trader of these products, whose annual consumption is constantly increasing. Globally, the share of seafood products is evenly divided between aquaculture and fisheries, while the EU market is dominated by fishery products that constitute about 75% of the market share. The marine fish production sector is the most economically important and has the largest turnover of €1,811 million, followed by the shellfish sector with €1,266 million, and the freshwater sector with €1,016 million. Production is mainly concentrated in four countries, namely, Spain (27%), France (18%), Italy (12%), and Greece (11%), which cover 62% of EU turnover. More than 80% of the companies in the aquaculture sector are micro-enterprises with less than ten employees, and it is estimated that there are about 15,000 of them. Rainbow trout and European seabass farming has been the most lucrative. Greece is a significant producer of gilthead seabream and European seabass with about 53% of the total EU production. France and Spain are the most important countries in production volume and value, employment, and the number of enterprises in the shellfish sector. Italy is the primary clam producer, covering 87% of EU production. The dominant freshwater species in aquaculture is trout, with a production volume of 53%, and the most important producers are Denmark (25%), Italy (21%), and France (20%). Carp is another important species mainly produced in Eastern Europe, especially in Hungary and Romania (Scientific, Technical and Economic Committee for Fisheries, STECF 2021). The EU is heavily dependent on imported seafood products, with imports accounting for about 60% of the total supply. Future demand for fish is expected to increase due to the growing population, rising incomes, and health benefits associated with fish consumption. Growing demand offers a unique opportunity to expand production in the EU, implying EU competitiveness in the global seafood market. In 2018, Asian countries dominated aquaculture production with about 90% of world production, while the EU contribution amounted to 1.0%–1.5% (Food and Agriculture Organization 2020). A prerequisite for the progress of the European aquaculture sector is to increase knowledge of existing aquaculture production. The European Green Deal considers aquaculture production to be a source of low-carbon protein for food and feed, which must comply with the strict EU and national requirements to ensure human, animal, and environmental health. The most important aspects of the environmental sustainability of EU aquaculture relate to assessing, monitoring, and limiting the impact of aquaculture activities on the environment (discharging organic matter from aquaculture farms into water), use of alien or locally absent species, ingredients of feeds for carnivorous fish (alternatives to wild fish), disease management, and the use of veterinary medicines and other substances with low environmental impact. Another increasingly important aspect refers to the welfare of animals in fish farming. The EU has the most extensive maritime space globally that will be increasingly used to generate clean, renewable energy by building offshore wind farms and using energy from wind, waves, tides, salinity gradient, and even algae (biofuel). Technology transforming these energy sources into electricity is developing rapidly, and marine renewables could become a major energy source in the EU by 2050. The goals of the strategy to harness the potential of offshore renewable energy that the European Commission issued in 2020 include gradually replacing fossil fuels with offshore renewables and creating industrial opportunities and green jobs across the continent. The marine renewables industry will have to scale up five times by 2030 and 25 times by 2050 to support the objectives of the Green Deal (European Commission 2020a). The European Commission is committed to supporting the value chain by investing in development and monitoring progress. Implementing the Maritime Spatial Planning Directive is key to anticipating change, avoiding conflicts, and finding synergies between different activities at sea (European Union 2014). Activities of the European Commission facilitate cross-border cooperation and the exchange of good practices on the issues of maritime spatial planning and marine renewables, particularly through the North Sea Energy Cooperation and various sea basin strategies.

Traditional sectors that contribute to the blue bioeconomy are marine biological resources, marine non-living resources, marine renewable energy, port activities, shipbuilding and ship repair, maritime transport, and coastal tourism. The marine biological resources sector refers to the collection of renewable biological resources and their conversion into food, feed, bioenergy, as well as bio-based products and their distribution along the supply chain. The marine non-living resources sector includes the extraction of crude petroleum and natural gas as well as other minerals, salt, gravel, sand, clay, and kaolin and has increased in recent decades. In addition to making a significant contribution to the EU 2050 Energy Strategy, marine renewable energy, which includes offshore wind and ocean energy, has great potential for sustainably generating economic growth and jobs, increasing the security of energy supply, and boosting competitiveness through technological innovation. Wave energy, tidal energy, salinity gradient energy, and ocean thermal energy conversion technologies are currently being developed to exploit a significant clean, renewable energy source. Offshore wind energy generated in offshore wind farms is currently the only commercial application of renewable marine energy. As the world's leading producer, Europe covers 90% of the world's total installed capacity. The leading producers are Germany, the Netherlands, Belgium, and Denmark. The port activities sector continues to play a vital role in trade, economic development, and job creation. The number of containers coming to European ports has increased by more than four times in the last 20 years, and the busiest container ports are Rotterdam, Antwerp, Hamburg, Amsterdam, and Algeciras (European Commission 2021c). More and more ports across the EU are seeking to reduce their environmental and climate impact by supporting green shipping fleets or acting as centres of clean energy, thus contributing to the goals of the Europe Green Deal. The EU shipbuilding sector is dynamic and globally competitive and specialises in building and repairing the most complex and technologically advanced civilian and naval ships, platforms, and hardware for maritime applications. Continuous investment in the sector through research, fostering innovation, and investing in a highly skilled workforce has positioned the EU as a world leader in producing high-tech, advanced maritime equipment and systems. Maritime transport plays an essential role in the EU economy and trade, estimated to account for around 80% of worldwide goods transportation and one-third of the intra-EU trade. International maritime shipping accounts for less than 3% of annual global GHG emissions and produces less exhaust gas for each tonne transported per kilometre than air or road transport, thus making a pivotal contribution to decarbonisation (European Commission 2021c). Due to the expected growth of the world economy, trade, and associated maritime transport demand, this sector's GHG emissions could increase from 50% to 250% by 2050 (European Commission 2021c) if adequate measures are not taken to reduce the impact on the environment and improve the energy efficiency of ships as well as the transition to alternative fuels. Coastal tourism is the largest sector of the blue bioeconomy in terms of gross value added and employment. However, since more than half of EU tourist accommodation is located in coastal areas and islands, these hotspots have recently been characterised by high building density and increasing numbers of tourists, which expands environmental footprints and negatively affects sustainable development. In 2017, the three most popular tourist destinations in the EU were the Canary Islands and Catalonia in Spain and the Adriatic coastal region in Croatia. Coastal tourism is particularly important for southern European countries, such as Spain, Portugal, Italy, Malta, and Greece, but also other coastal countries such as Croatia, Bulgaria, Romania, and the Netherlands. Due to its contribution to the global economy, *cruise tourism* is a significant and growing segment within coastal tourism that has witnessed a 53% growth in Europe over the last decade. Europe is the largest cruise ship builder and the second most popular cruise destination in the world (European Commission 2021c).

Emerging sectors include ocean energy, blue bioeconomy and biotechnology, desalination, and marine mineral resources. Ocean energy includes various renewable energy technologies that use wind, ocean energy (tidal power, ocean thermal energy conversion, salinity gradient), floating solar photovoltaic energy, and renewable hydrogen. Wind exploitation represents the most advanced sector while other technologies are at the early stages of development. The ambitious growth of the sector is based on the extremely high energy potential of Europe's sea basins and the EU's globally leading position in the sector. The EU's blue bioeconomy and biotechnology include marine organisms that do not have a traditional commercial application and are used as biomass. Algae (macro- and micro-), bacteria, fungi, and invertebrates are important marine resources used as raw materials for various commercial applications, including food and food supplements, feeds, cosmetics, fertilisers, and innovative biomaterials for bioremediation or biofuels. These represent important resources and help highlight the innovative nature and potential of the sector given a number of EU priorities such as carbon neutrality; innovative, healthy, and sustainable food systems; and a sustainable and circular economy. Spain, France, Ireland, and Norway are European countries with the largest number of macroalgae production companies. Activities related to the macroalgae industry represent an important cultural heritage and a significant source of income for some coastal and rural communities. Algae represent an incredibly versatile material, with potentially new applications in various economic sectors, such as pharmaceutical, food, textile, chemical, cosmetics, packaging, agricultural, construction, and biofuel. Seaweed aquaculture, especially in combination with shellfish aquaculture, will ensure healthy food on the market and contribute to the ecosystem through carbon sequestration, ocean habitat restoration, and strengthening coastal ecosystem resilience. New jobs and sustainable growth of coastal regions result from EU activities and investments in the blue bioeconomy. Algae production in the EU, although limited, is on the rise. The improvement of this production method relies on overcoming technological constraints to reduce infrastructure and logistics costs and increase biomass yields. Special attention is paid to biorefineries that process algae to increase environmental sustainability and economic feasibility compared to conventional industrial processes. Different algal biomass conversion pathways are researched for the use, extraction, and valorisation of value-added products. This cultivation method offers advantages in maritime space management and increasing production capacity. Integrated Multi-Trophic Aquaculture is a way to increase the ecological and economic sustainability of the production of all cultivated species. The approach is based on the co-cultivation of species from different trophic levels to potentially reduce the nutrients and organic matter inputs from aquaculture. The cultivation and harvesting of less exploited groups of organisms (e.g., sea urchins or starfish) is being researched as a way to reduce the pressure on natural resources in certain areas and increase the diversification of aquaculture to low trophic levels. However, these activities are still at a very early stage of development in Europe. Desalination is an alternative water supply that can lessen the growing pressure on freshwater resources. Currently, desalination technology is used to overcome water scarcity in areas where freshwater resources are limited, such as large coastal cities, islands, and offshore industrial plants where seawater cannot be used due to high salinity. Many EU regions will face severe water scarcity by 2050. These areas include the coastal Mediterranean regions and other regions in France, Germany, Hungary, Italy, Romania, and Bulgaria. In the long run, the demand for desalination and other water management solutions, such as water reuse, is expected to reduce the impact of climate change on freshwater availability. More than 75% of desalination capacity in coastal areas is located in the Mediterranean Sea basin, with Spain being the leading country with 65% of the desalination capacity in the EU. Spain is followed by Italy (7.5%), France (3.5%), Cyprus (3.4%), Malta (2.9%), and Greece (2.8%). Desalination plants located in Northern European countries such as Germany (4%), the Netherlands (3.8%), Belgium (1.9%), and Ireland (1.1%) are mainly associated with the production of drinking and industrial water. European companies are among the leaders when it comes to patents, research, and development related to desalination powered by renewable energy sources such as wave or wind energy. Marine mineral resources include marine aggregates (e.g., sand and gravel), other minerals and metals in/on the seabed (e.g., manganese, titanium, copper, zinc, and cobalt), and chemical elements dissolved in seawater (e.g., salt and potassium) (European Commission 2021c).

The European Commission presented an action plan (European Commission 2020b) tackling issues of secure access to sustainable raw materials, given that ensuring reliable access to raw materials has become an important strategic factor in the EU's competitiveness. There are currently no commercial seabed mining projects in the area under the national jurisdiction of EU member states. Although scientists claim the loss of biodiversity due to deep-sea mining is likely inevitable and lasting, the EU will continue to fund research into the impact of deep-sea mining activities and environmentally friendly technologies. These scientific discoveries have raised public awareness and stimulated political debate within the EU and beyond. The European Parliament adopted a resolution on international oceans governance in January 2018 (European Parliament 2018), calling for a moratorium on commercial permits for submarine exploitation until the risks to the environment are fully understood. The potential damage of mining impacts requires developing environmentally friendly technologies that can limit the generation of plumes and other adverse environmental impacts during mining (sediment disturbance, release of toxic compounds, light and noise generation, thermal pollution) as well as developing tailored policies. The EU has funded a series of studies and projects to increase knowledge of deep-sea marine mineral resources and ecosystems to understand better the potentially harmful effects of mining on the environment.

Conclusion

The concept of a bioeconomy is based on the assumption that biomass is currently underutilised as many waste streams are not used optimally and that the biomass potential can be increased by increasing current yields and productivity of land, introducing new species and new and improved extraction and processing technologies. The bioeconomy can increase the value generated from biomass and stimulate value chains by cascading the use of biomass and reusing waste materials. The development of the bioeconomy is expected to contribute to a broader range of valuable by-products that could reduce the pressure on food and feed, and the EU strategy also advocates the principle of cascading use of biomass and waste streams (El-Chichakli *et al.* 2016). However, proper implementation of the cascading use of biomass depends on regional, local economic and technological circumstances, maintaining the necessary carbon stock in the soil, and the state of the soil and ecosystem.

Achieving sustainable development in the bioeconomy poses many challenges, such as tackling climate change, sustainably managing natural resources, and deciding between different biomass uses while ensuring social inclusion. In addition, sustainable production of renewables is necessary to convert these resources and waste streams into value-added products, such as food, feed, bio-based products, and bioenergy.

Advances in natural sciences and engineering play a central role in developing the bioeconomy as they can be applied in various sectors of the economy, from the processing of bioresources to the establishment of new value chains (Bugge *et al.* 2016), which is why the bioeconomy is not a "new" sector by itself, but a set of intersecting value chains of different sectors covering agriculture, forestry, fisheries, food processing, and parts of the energy, chemical, and biotechnology sectors (Wessler and von Braun 2017).

The goals of the cross-sectoral bioeconomy are food security and food supply, sustainable management of natural resources, reduction of dependence on non-renewable and unsustainable resources, mitigation of and adaptation to climate change, as well as strengthening of competitiveness and job creation. To support these goals, it is necessary to strengthen and increase the sectors based on biological resources, increase investment and expand the market, and implement local bioeconomy strategies across Europe, considering objective constraints. The primary goal of cross-sectoral and social integration is focused on local, new value chains. It is based on reducing dependence on imported raw materials, replacing raw materials of fossil origin, creating an efficient material cycle system, applying local knowledge in innovative partnerships, strengthening local socio-economic relations, and relying on the community to reduce the environmental burden (European Commission 2018).

Compromises and synergies in the transition to the bioeconomy will affect the production, supply, and transport of biomass and areas of skilled labour, land use, new waste streams, niche markets, and national funds. The development of the bioeconomy largely depends on highly skilled labour with a wide range of knowledge and technological expertise capable of adapting to innovation and structural change (Schmid *et al.* 2012). These should be highly qualified individuals with interdisciplinary education, strong entrepreneurial skills, and experience in the biological sciences, natural resources, agronomy, biotechnology, or bioengineering. The transition to the bioeconomy must be viewed through the perspective of overall environmental impacts in line with the goals of sustainability and the protection of biodiversity. It is necessary to assess the environmental impact of sustainable production and biomass use since the fact that a product is bio-based does not in itself guarantee its viability or quality. Therefore, it is necessary to consider the GHG balance, resource efficiency with regard to energy and water consumption, land use and biomass, ecosystem services maintenance and its biological and landscape dimensions, the management of waste and by-products, and the possibility of recycling end products. The need to increase crop productivity could lead to increased use of fertilisers and pesticides, with additional water and soil pollution problems. The bioeconomy could also affect water scarcity in some parts of the world due to additional pressure on water demand. Increased land demand for food and non-food crops could result in monocultures, negative environmental impacts, and increased pressure on natural habitats and biodiversity. The future development of the bioeconomy could also lead to a radical redesign of products and processes, generate demand for new skills, and create new markets, which might simultaneously lead to the obsolescence of certain products, processes, and skills. A bioeconomy based on large industrial plants may result in concentration and intensification of international trade, with an uneven geographical and social distribution of costs and benefits and a net loss of jobs. The food and agricultural sectors face multiple challenges that include meeting growing global demand for food and agricultural products due to the growing population, rising income, and related changes in eating habits, as well as growing demand for biomass to meet the needs of the energy and industrial raw materials' sectors. Meeting these requirements will increase the pressure on the agri-food sector to supply food and raw materials from scarce natural resources while protecting the environment in the context of climate change. Facing these challenges sustainably will require developing new products and improving existing technologies and practices. The growing demand for agricultural biomass related to the development of the bioeconomy must be seen in the context of limited arable land and shrinking agricultural land in many countries, which entails a high risk of increased competition between food supply and non-food biomass production (European Commission 2012). The excessive emphasis on alternative biomass use would shift the focus from food production and landscape preservation and management. In contrast, the necessary increase in biomass production would rely on increased productivity and increased resource efficiency to ensure the use of biological resources within their sustainability thresholds. Risk reduction and growth of the bioeconomy require close cooperation of all stakeholders (Philp and Winickoff 2017) for a coherent and integrated policy approach to achieve more sustainable outcomes for agriculture and the food system. The main challenge for decision-makers in the private and public sectors is determining the most cost-effective biological resource to meet food, feed, fuel, and fibre needs. While reducing GHG emissions is a key driver for developing the bioeconomy, there are concerns about the overall GHG savings resulting from the production of raw materials, land-use change, and bioenergy conversion. There is a lack of concrete empirical evidence on the overall net economic, environmental, and social impacts of the bioeconomy on sectors, so better monitoring and evaluation are needed. In addition, it is necessary to overcome regulatory constraints, address transport logistics, facilitate market opportunities, and develop indicators for monitoring and assessing the impact. Therefore, the effects of land-use change on biodiversity and local, regional, and global socio-economic systems are considered. When measuring the ecological performance of products and bioeconomy processes, it is necessary to assess the life cycle of products that includes the entire supply chain and its interconnected processes, namely the production and processing of biomass and production and use of final products. Such assessments may reveal that certain renewable energy production processes may be inefficient or costly, given their market requirements (Sevigne-Itoiz *et al.* 2021). The updated EU Bioeconomy Strategy (European Commission 2018) advocates building an internationally harmonised system for monitoring economic, environmental, and social progress towards sustainable bioeconomy at the EU level and stresses the contribution of life-cycle assessment (LCA) in developing methodological standards for bio-based products. LCA can also contribute to determining the most efficient use of biomass and biowaste. Another approach refers to measuring the share of renewable bio-content in products and services in the economy. However, this approach ignores sustainability issues, such as the origin of resources and how their production and use are linked to more sustainable development. On the other hand, outcome-based approaches, such as reduced carbon emissions and the sustainability of water, soil, and biodiversity improvements, are more promising, even though they are the most demanding in terms of data and methodological challenges. Since the increased demand for bioresources represents great pressure on limited resources and biomass, which is a direct threat to sustainability, the use of new technologies can reduce the negative impact of the excessive use of biomass, and the proper implementation of innovation can mitigate the adverse effects on the ecosystem (Gawel *et al.* 2019). Innovations include substitute products based on the use of bioresources instead of fossil products, new processes that promote the bio-based production and value chains, new bio-based materials with new characteristics, as well as new types of behaviour, interaction, and collaboration of users with other stakeholders. The transition to a bioeconomy is a complex process that requires well-designed regulation, a resilient economy, and an innovation-driven industry. Many national strategies emphasise the environmental benefits of innovation if implemented gradually, considering that innovation improves sustainability (D'Amato *et al.* 2017). The EU strategy places special emphasis on strengthening the dialogue between researchers in the field of bioeconomy and decision-makers to ensure that public research provides a solid scientific basis for responsible political decisions while taking into account the existing communication noise related to the knowledge and information transfer between innovators, researchers, biotechnologists, and agricultural communities. Bioeconomy growth requires an integrated, coherent approach with close cooperation and coordination between businesses, policymakers, civil society, and scientists for a coherent and integrated policy approach to achieve more sustainable outcomes for agriculture and the food system (Food and Agriculture Organization of the United Nations 2016).

Bibliography

Bugge, M. M., Hansen, T., and Klitkou, A., 2016. What is the bioeconomy? A review of the literature. *Sustainability*, 8 (7), 691. Available from: www.mdpi.com/2071-1050/8/7/691/htm

Carus, M., 2017. The bioeconomy is much more than a circular economy. Available from: www.brain-biotech.com/blickwinkel/circular/the-bioeconomy-is-much-more-than-a-circular-economy

D'Amato, D., Droste, N., Allen, B., Kettunen, M., Laehtinen, K., Korhonen, J., Leskinen, P., Matthies, B. D., and Toppinen, A., 2017. Green, circular, bio economy: A comparative analysis of sustainability avenues. *Journal of Cleaner Production*, 168, 716–734. Available from: www.sciencedirect.com/science/article/pii/S0959652617320425

Diakosavvas, D., and Frezal, C., 2019. *Bioeconomy and the Sustainability of the Agriculture and Food System: Opportunities and Policy Challenges* (OECD Food, Agriculture and Fisheries Papers, No. 136). Paris: OECD Publishing. Available from: http://dx.doi.org/10.1787/d0ad045d-en

El-Chichakli, B., Von Braun, J., Lang, C., and Barben, D., 2016. Policy: Five cornerstones of a global bioeconomy. *Nature*, 535, 221–223. Available from: www.nature.com/articles/535221a

European Commission, 1998. Communication from the commission to the council and the European Parliament on a forestry strategy for the European Union (COM 1998/649 Final). Available from: https://eur-lex.europa.eu/legal-content/EN/TXT/PDF/?uri=CELEX:51998DC0649&from=NL

European Commission, 2003. Natura 2000 and forests "Challenges and opportunities" – Interpretation guide. Available from: https://ec.europa.eu/environment/nature/info/pubs/docs/nat2000/n2kforest_en.pdf

European Commission, 2010. Communication from the Commission Europe 2020: A strategy for smart, sustainable and inclusive growth (COM(2010)2020). Available from: https://ec.europa.eu/eu2020/pdf/COMPLET%20EN%20BARROSO%20%20%20007%20-%20Europe%202020%20-%20EN%20version.pdf

European Commission, 2012. Assessment of resource efficiency in the food cycle, final report. Available from: https://ec.europa.eu/environment/eussd/pdf/foodcycle_Final%20report_Dec%202012.pdf

European Commission, 2013. Communication from the Commission to the European Parliament, the Council, the European Economic and Social Committee and the Committee of the Regions. A new EU Forest Strategy: For forests and the forest-based sector. Available from: https://eur-lex.europa.eu/resource.html?uri=cellar:21b27c38-21fb-11e3-8d1c-01aa75ed71a1.0022.01/DOC_1&zformat=PDF

European Commission, 2015. Communication from the Commission to the European Parliament, the Council, the European economic and Social Committee of the Regions – Closing the loop – An EU action plan for the circular economy. Available from: https://eur-lex.europa.eu/resource.html?uri=cellar:8a8ef5e8-99a0-11e5-b3b701aa75ed71a1.0012.02/DOC_1&format=PDF

European Commission, 2018. Communication from the Commission to the European Parliament, the Council, the European Economic and Social Committee and the Committee of the Regions: A sustainable bioeconomy for Europe: Strengthening the connection between economy, society and the environment, COM/2018/673 final. Available from: https://eur-lex.europa.eu/legal-content/EN/TXT/?uri=CELEX%3A52018DC0673

European Commission, 2019. Bioeconomy: The European way to use our natural resources. Action plan 2018. Directorate-General for Research and Innovation, 1–26. Available from: https://op.europa.eu/en/publication-detail/-/publication/775a2dc7-2a8b-11e9-8d04–01aa75ed71a1

European Commission, 2020a. Communication from the Commission to the European Parliament, the Council, the European Economic and Social Committee and the Committee of the regions. An EU Strategy to harness the potential of offshore renewable energy for a climate neutral future, COM/2020/741. Available from: https://eur-lex.europa.eu/legal-content/EN/TXT/?uri=COM%3A2020%3A741%3AFIN

European Commission, 2020b. Communication from the Commission to the European Parliament, the Council, the European Economic and Social Committee and the Committee of the Regions. Critical raw materials resilience: Charting a path towards greater security and sustainability, COM/2020/474 final. Available from: https://eur-lex.europa.eu/legal-content/EN/TXT/?uri=CELEX:52020DC0474

European Commission, 2021a. Communication from the Commission to the European Parliament, the Council, the European Economic and Social Committee and the Committee of the regions: New EU forest strategy for 2030. Available from: https://eur-lex.europa.eu/legal-content/EN/TXT/?uri=CELEX:52021DC0572

European Commission, 2021b. Communication from the Commission to the European Parliament, the Council, the European Economic and Social Committee and the Committee of the Regions. New approach for a sustainable blue economy in the EU transforming the EU's blue economy for a sustainable future, COM (2021) 240 final. Available from: https://eur-lex.europa.eu/legal-content/EN/TXT/PDF/?uri=CELEX:52021DC0240&qid=1624869846426&from=EN

European Commission, 2021c. *The EU Blue Economy Report 2021*. Luxembourg: Publications Office of the European Union. Available from: https://op.europa.eu/en/publication-detail/-/publication/0b0c5bfd-c737-11eb-a925–01aa75ed71a1

European Commission, n.d. The Pan-European ministerial conference on the protection of forests in Europe. Available from: https://ec.europa.eu/environment/forests/mcpfe.htm

European Environment Agency, 2016. Circular economy in Europe – Developing the knowledge base (EEA Report 2/2016). Available from: www.eea.europa.eu/publications/circular-economy-in-europe

European Parliament, 2018. European Parliament resolution of 16 January 2018 on international ocean governance: An agenda for the future of our oceans in the context of the 2030 SDGs (2017/2055(INI)). Available from: https://eur-lex.europa.eu/legal-content/EN/TXT/?uri=CELEX%3A52018IP0004

European Union, 2014. Directive 2014/89/EU of the European Parliament and of the Council of 23 July 2014 establishing a framework for maritime spatial planning. *Official Journal of the European Union L257/135*. Available from: http://data.europe.eu/eli/dir/2014/89/oj

Food and Agriculture Organization, 2020. *The state of world fisheries and aquaculture 2020. Sustainability in Action*. Rome: FAO. Available from: https://www.fao.org/documents/card/en/c/ca9229en/

Food and Agriculture Organization of the United Nations, 2016. The State of Food and Agriculture: Climate change, agriculture and food security. Available from: www.fao.org/3/i6030e/i6030e.pdf

Food and Agriculture Organization of the United Nations, 2018. Assessing the contribution of bioeconomy to countries' economy. A brief review of national frameworks. Available from: www.fao.org/3/I9580EN/i9580en.pdf

Forest Europe, 2015. State of Europe's forests 2015. Available from: www.foresteurope.org/docs/fullsoef2015.pdf

Forest Europe, 2019. Forests for human health: Challenges and opportunities. Vienna Workshop Report, 8–9 April. Available from: https://foresteurope.org/wp-content/uploads/2016/08/Workshop-report-HHWB_ver2.pdf

Gawel, E., Pannicke, N., and Hagemann, N., 2019. A path transition towards a bioeconomy – The crucial role of sustainability. *Sustainability*, 11 (11), 3005. Available from: www.mdpi.com/2071-1050/11/11/3005

Intergovernmental Oceanographic Commission of UNESCO, 2021. MSPglobal: International guide on marine/maritime spatial planning. Available from: https://unesdoc.unesco.org/ark:/48223/pf0000379196

Mapping and Assessment of Ecosystems and Their Services, 2013. *An Analytical Framework for Ecosystem Assessments under Action 5 of the EU Biodiversity Strategy to 2020*. Luxembourg: Publications office of the European Union. Available from: https://ec.europa.eu/environment/nature/knowledge/ecosystem_assessment/pdf/MAESWorkingPaper2013.pdf

Millennium Ecosystem Assessment, 2005. *Ecosystems and Human Well-being: Synthesis*. Washington, DC: Island Press. Available from: www.millenniumassessment.org/documents/document.356.aspx.pdf

Ontl, T. A., and Schulte, L. A., 2012. Soil carbon storage. *Nature Education Knowledge*, 3 (10), 35. Available from: www.nature.com/scitable/knowledge/library/soil-carbon-storage-84223790/

Organisation for Economic Co-operation and Development, 2018. Meeting policy challenges for a sustainable bioeconomy. Available from: https://doi.org/10.1787/9789264292345-en

Organisation for Economic Co-operation and Development, 2019. Bio-economy and the sustainability of the agriculture and food system: Opportunities and policy challenges. Available from: www.oecd.org/officialdocuments/publicdisplaydocumentpdf/?cote=COM/TAD/CA/ENV/EPOC(2018)15/FINAL&docLanguage=En

Philp, J., and Winickoff, D. E., 2017. Clusters in industrial biotechnology and bioeconomy: The roles of the public sector. *Science & Society*, 35 (8), 682–686. Available from: https://doi.org/10.1016/j.tibtech.2017.04.004

Priefer, C., Jörissen, J., and Frör, O., 2017. Pathways to shape the bioeconomy. *Resources*, 6 (1), 10. Available from: https://doi.org/10.3390/resources6010010

Schmid, O., Padel, S., and Levidow, L., 2012. The bio-economy concept and knowledge base in a public goods and farmer perspective. *Bio-based and Applied Economics*, 1 (1), 47–63. Available from: www.researchgate.net/publication/238596455_The_Bio-Economy_Concept_and_Knowledge_Base_in_a_Public_Goods_and_Farmer_Perspective

Scientific, Technical and Economic Committee for Fisheries, 2021. *The EU Aquaculture Sector – Economic Report 2020 (STECF-20–12)*, In R. Nielsen, J. Guillen Garcia and J. Virtanen, eds. Luxembourg: Publications Office of the European Union. Available from: https://doi.org/10.2760/441510

Sevigne-Itoiz, E., Mwabonje, O., Panoutsou, C., and Woods, C., 2021. Life cycle assessment (LCA): Informing the development of a sustainable circular bioeconomy? *Philosophical Transactions of the Royal Society A*, 379, 20200352. Available from: http://doi.org/10.1098/rsta.2020.0352

Staffas, L., McCormick, K., and Gustavsson, M., 2013. A global overview of bio-economy strategies and visions. *Swedish Knowledge Centre for Renewable Transportation Fuels (f3 Centre)*. Available from: http://f3centre.se/sites/default/files/f3_report_2013-6_global_bioeconomy_overview_130410.pdf

Standing Committee on Agricultural Research-Agriculture Knowledge and Innovation System, 2017. Policy brief on new approaches on agricultural education systems. Available from: https://ec.europa.eu/eip/agriculture/sites/default/files/policy_brief_on_education_systems_scar_akis_06102017.pdf

United Nations, 2015a. Resolution adopted by the General Assembly on 25 September 2015. Transforming our world: The 2030 Agenda for Sustainable Development (A/RES/70/1). Available from: www.un.org/ga/search/view_doc.asp?symbol=A/RES/70/1&Lang=E

United Nations, 2015b. Paris Agreement. Available from: http://unfccc.int/files/essential_background/convention/application/pdf/english_paris_agreement.pdf

von Braun, J., 2018. Global institutions: Governance reform for food, nutrition, and agriculture. In *2018 Global Food Policy Report*. Washington, DC: International Food Policy Research Institute (IFPRI), 62–71. Available from: https://doi.org/10.2499/9780896292970_08

Wessler, J., and von Braun, J., 2017. Measuring the bioeconomy: Economics and policies. *Annual Review of Resource Economics*, 9 (1), 275–298. Available from: https://papers.ssrn.com/sol3/papers.cfm?abstract_id=3060479

Wolfslehner, B., Linser, S., Pülzl, H., Bastrup-Birk, A., Camia, A., & Marchetti, M., 2016. Forest bioeconomy – A new scope for sustainability indicators, from science to policy 4, European Forest Institute. Available from: https://efi.int/sites/default/files/files/publication-bank/2018/efi_fstp_4_2016.pdf

4 New emerging sectors

Anamarija Farkaš

Introduction

The challenges and tasks of the bioeconomy refer to balancing the goals and needs of economic development with conserving biological diversity and preserving quality in all terrestrial and aquatic ecosystems. The growing awareness of sustainability and the necessity of reducing climate change stimulate the development of bio-based industries and bio-based innovation (Vehvilainen *et al.* 2021). *New emerging sectors* focus on the knowledge-based bioeconomy and biotechnological innovation, management, and economic models to enable commercial success that entails environmental protection and achieving system sustainability (Figure 4.1).

While the bioeconomy is currently included in various disciplines, it plays a vital role in natural and engineering sciences, as shown by the bibliometric

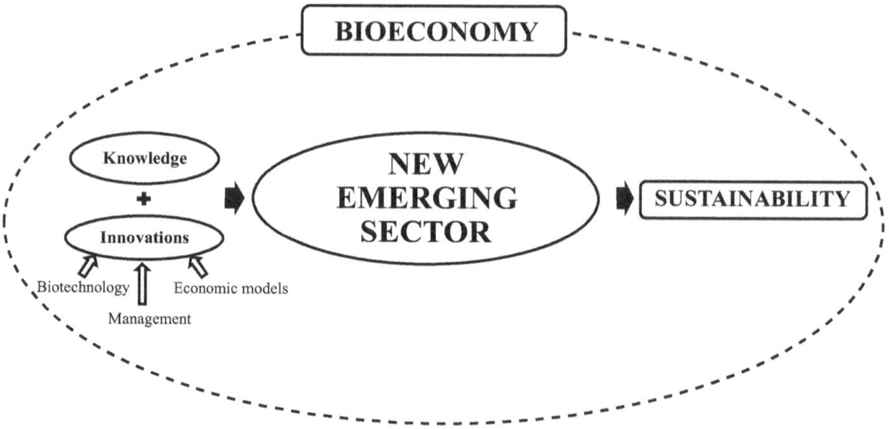

Figure 4.1 New emerging sectors that focus on the knowledge-based bioeconomy and innovations

Source: Author

DOI: 10.4324/9781003223733-5

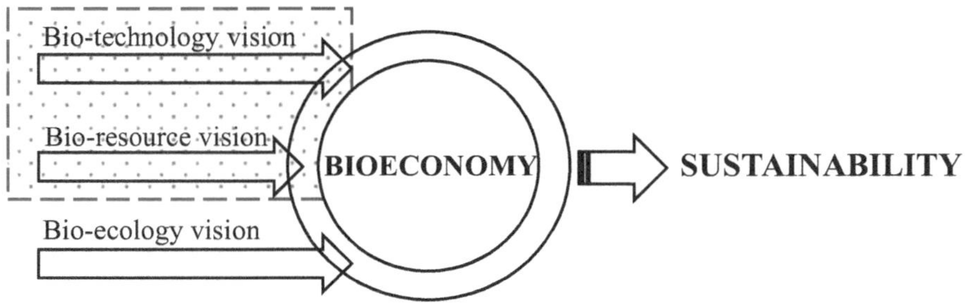

Figure 4.2 A vision of the bioeconomy with the focus on sustainability according to bibliometric analysis in research science

Source: Author

analysis conducted in the research community. A literature review in the scientific and research communities resulted in the identification of three visions of the bioeconomy: (1) the biotechnology vision, which emphasises the importance of biotechnology research and its application and commercialisation in different sectors of the economy; (2) the bioresource vision relating to the processing and upgrading of biological raw materials and the establishment of new value chains; and (3) the bioecology vision, which highlights sustainability and ecological processes that optimise the use of energy and nutrients, promote biodiversity, and avoid monocultures and soil degradation. The biotechnology and bioresource visions partially overlap, which could lead to a complementary strategy in terms of the possible application of biotechnology to bioresources (Figure 4.2) (Bugge *et al.* 2016).

Cellular and biomolecular processes for the development of technologies and products that will contribute to the well-being of humankind are applied due to the connection between natural and engineering sciences in the biotechnology (Biotechnology Innovation Organization 2021).

Biotechnology is used in various industrial sectors, such as healthcare and pharmaceuticals, animal health, textiles, chemicals, plastic, paper, fuel, food, and food processing. In *healthcare and pharmaceutical* applications, biotechnology has led to the discovery and development of advanced medicines, therapies, diagnostics, and vaccines. Biotechnological innovations have helped develop novel drugs for patients with growth diseases, metabolic diseases, multiple sclerosis, rheumatoid arthritis, cancer, and Alzheimer's disease. In *agriculture, livestock, veterinary products, and aquaculture*, innovations have made it possible to improve animal feed, produce novel vaccines for livestock, and improve diagnostics for disease detection. Biotechnology has also enabled the use of enzymes for more efficient food processing and improved plant breeding. In *industrial processes and manufacturing*, innovations have allowed the use of enzymes in the production of detergents, pulp and paper, textiles, and biomass. By using fermentation and enzyme biocatalysis instead of traditional chemical synthesis, higher process efficiency can be obtained, decreasing energy and water consumption as well as reducing toxic waste (Biotechnology Innovation Organization 2021).

In many countries, the application of modern-day industrial biotechnology based on innovation enables the growth of the economy and the creation of new jobs while supporting sustainable development, human health, and environmental protection. Despite the economic challenges, industrial biotechnology is expected to grow. The driver of innovation in industrial biotechnology could also be the need to act given increasing climate change (Batchelor 2012).

New emerging sectors in the field of biotechnology have contributed to the modernisation of the industry because modern biotechnology has made significant advances in technology and innovative products. Seven key factors enable innovation in biotechnology (Figure 4.3).

Creating conditions for developing biotechnological innovations is almost impossible without highly qualified and technically competent *human resources*. Combined with having adequate, educated, and technically proficient levels of human capital, *research and development (R&D) infrastructure and capacity* is essential to successfully foster innovation and activities in high-tech sectors, including biotechnology. Biotechnological R&D would be almost impossible without laboratories and clinical research facilities. *Intellectual property rights* (including patents and regulatory data protection) are essential for biotech and biopharmaceutical innovations. For biopharmaceutical and non-pharmaceutical biological products and technologies, intellectual property rights encourage and support the R&D of novel technologies and bio-based products. *Well-established regulations* (legislation) in a given country or region play a significant role in

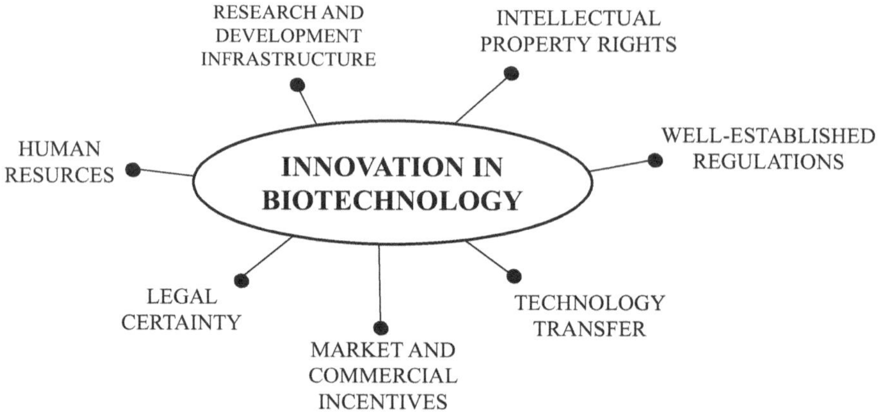

Figure 4.3 Seven key factors enabling innovation in biotechnology

Source: Author

shaping incentives for innovation and establishing appropriate levels of quality and safety for biotech products, especially biopharmaceuticals. They also create conditions for producing and selling high-quality products and technologies. *Technology transfer* is a crucial mechanism for commercialising and transferring research from public and governmental bodies to private entities and private-to-private entities to develop usable and commercially available technologies. *Market and commercial incentives* range from general R&D incentives to specific policies aimed at biotech sectors, such as pricing and reimbursement policies for biopharmaceuticals. For the biopharmaceutical sector, incentives determined by pricing and reimbursement systems for medicines and health technologies can substantially impact commercial and market incentives for innovation in health and biotech R&D. *Legal certainty* (including what relates to the rule of law and the rule of law within a business context) is crucial to commercialisation and business activities (Building the Bioeconomy 2018).

Start-ups and small and medium-sized enterprises (SME) are currently at the forefront of new emerging sectors. However, they struggle to develop and survive in a well-established fossil fuel market by anticipating financial risks and securing long-term investments. As consumer habits are changing towards more sustainable practices, it is imperative for companies to start radical changes to stay in business (Plug and Play 2021).

The bio-based industry is part of the economy formed by companies that use biological resources as inputs to produce materials, products, and services. In contrast, bio-based innovation is a new concept, technology, process, material, or product based on the use and transformation of biological input (Vehvilainen *et al.* 2021). The characteristics of the critical parameters of the bio-based industry (bioresources and bio-based innovation) are shown in Table 4.1.

Developing a sustainable bio-based industry contributes to creating competitiveness, innovation, and new jobs. In contrast, the so-called green innovations improve human health and quality of life, reduce greenhouse gas (GHG)

Table 4.1 Key parameters of the bio-based industry

Bio-based Industry	
Bioresources	Biomass extracted from the natural environment and purpose-grown biomass (e.g., from agriculture and forestry, fisheries, and aquaculture), as well as various forms of biological waste, by-products, and residues
Bio-innovation	The benefits of bio-based innovation include one or more effects of biological production: Increased energy or material efficiency of the production process New properties of the produced material or product The ability to use and valorise waste The elimination of pollution

emissions and water use in production, use natural resources sustainably, and, most importantly, improve soil, water, and forest biodiversity. In addition, the obtained products are healthier than those based on fossil resources, while bio-based circular production processes contribute to a significant reduction in waste and plastic pollution.

New emerging sectors will significantly change existing industries based on redesigned business models that will offer new innovation-based capabilities but also reduce new types of risks caused by the Covid-19 pandemic and the global environmental problem that is climate change.

Each country should synergise research and innovation to improve and upgrade the use of its bioresources and technology and create new value chains at the local level instead of exporting bioresources for upgrading elsewhere, namely in other countries (Bugge *et al.* 2016). However, it is also crucial to raise citizens' awareness and understanding of the bioeconomy and empower them to become actors of change.

This chapter analyses *specific emerging sectors* that significantly impact the bioeconomy, such as bioenergy, bio-based chemistry, biochemistry, and biopharmaceutical sectors. The analysis of individual sectors includes concrete examples of new research and bio-based industries, as well as the challenges that some sectors face in achieving their goals in addition to their common goal relating to sustainability and climate neutrality.

The bioenergy sector

The world is still excessively dependent on fossil fuels to meet its energy needs. Energy derived from fossil fuels accounts for 81.5% of total energy consumption in the United States, making fossil fuels the primary energy source. However, the scarcity of raw materials and the worrying increase of carbon dioxide and other GHGs in the atmosphere caused by fossil fuels emphasise the need to use energy derived from biological resources and reduce the dependence on conventional energy globally. It is estimated that current global crude oil reserves are sufficient to meet global consumption by 2050, while coal and gas shortages could occur in the next 40 to 50 years. India and China, the countries with the highest GHG emissions, are taking decisive action to reduce those emissions. In Japan, the use of wood as a biofuel contributes to the increase in the share of bioenergy use in the whole Asia-Pacific region (Fortune Business Insights 2020b).

Bioenergy, which covers approximately 10% of the global total energy needs, is energy derived from biofuels directly or indirectly produced from organic material or biomass, including plant material and animal waste. Biofuels can also be obtained from wood, crops, or waste materials using more advanced technologies. Solid biofuels (see Table 4.2.) include organic, non-fossil materials of biological origin that can be used as fuel for heat production or electricity generation using existing or innovative technologies. Primary biofuels (firewood, wood chips, and pellets) are used in unprocessed form, primarily

Table 4.2 Types of solid fuels for bioenergy production and positive and negative environmental impacts

	Types	*Environmental impact*	
		Positive	*Negative*
Solid Biofuels	**Wood** (firewood, charcoal, wood chips, pellets, and sawdust)	- Less environmental impact than fossil fuels – Contributes to the reduction of global warming	– Particulate and polycyclic aromatic hydrocarbon emissions – Impact on human and animal health
	Municipal waste (almost everything that people throw away every day (food waste, plant waste, etc.)	– Reduces waste	– If controlled waste incineration techniques are not used, hazardous amounts of contaminants and toxic gases may be generated
	Animal waste (dried animal dung)	– Cheap – Available	– The incineration process is more difficult than the municipal waste incineration process – Waste contains many pollutants (higher levels of dioxins and chlorophenols) – Great danger to human health if incinerated indoors with limited ventilation
	Energy crops (crops grown exclusively for incineration in the form of pellets, briquettes [*Miscanthus*])	– Cheap production of clean energy, with no CO_2 emissions during combustion	– Lack of moisture in the soil is prevented by irrigation which entails higher water consumption, use of fertilisers, herbicides

for heating, cooking, or electricity generation. In contrast, secondary biofuels, derived from biomass and animal or vegetable waste, include liquid biofuels such as bioethanol, biodiesel, and so on which are used in vehicles and industrial processes. Liquid biofuels obtained through thermochemical, chemical, biochemical, and mechanical processes (see Figure 4.4), including bioethanol produced from crops containing sugar and starch and biodiesel from oilseeds, are referred to as "first-generation biofuels". However, these use only a part of the energy potentially available in biomass. Today, about 60% of bioethanol is produced from maise, 25% from sugar cane, 7% from molasses, 4% from wheat,

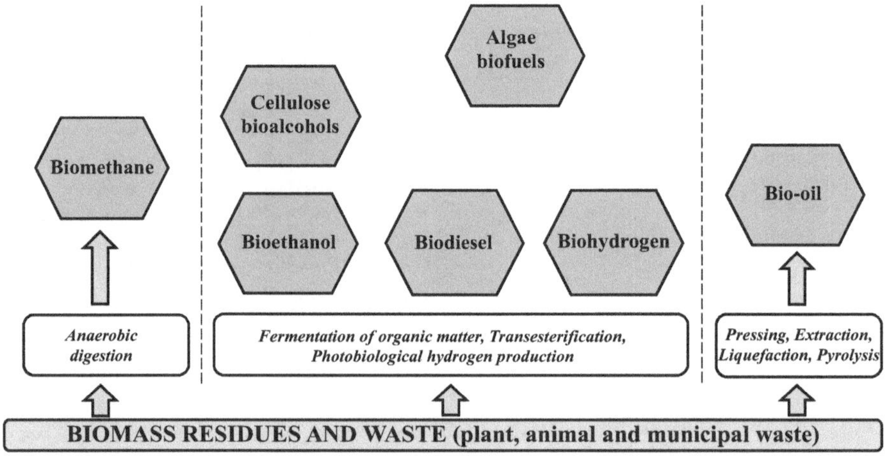

Figure 4.4 Liquid biofuels and technological procedures for their production

Source: Author

and the rest from other grains or sugar beet. About 77% of biodiesel is obtained from vegetable oils (30% soybean oil, 25% palm oil, 18% rapeseed oil) or waste cooking oils (22%).

"Second-generation biofuels" refer to liquid biofuels obtained by technological processes from components of plant substances such as cellulose, hemicellulose, and lignin. In contrast, the next generation of biofuels, which refers to securing bioenergy from algae materials, is the subject of new research and new (innovative) technologies that would be applicable globally. In addition to those liquid biofuels in Figure 4.4, there are several other liquid biofuels currently in use, such as Bio-MTBE (methyl-tert-butyl ether) as a fuel produced from biomethanol, Bio-ETBE (ethyl-tert-butyl ether) as a fuel produced from bioethanol, Bio-DME (dimethylether) produced from biomass, and HTU (Hydrothermal Upgrading) diesel obtained from wet mass of animal origin (Green Facts 2009; Lee *et al.* 2019; OECD/FAO 2019; European Commission 2019b, 2019c).

As one of the largest producers of liquid biofuels, North America is constantly expanding its bioenergy market share by significantly increasing bioethanol production. European Union (EU) countries are increasingly using bioenergy instead of fossil fuels, which is contributing to the EU's accelerated transition to renewable energy in the coming decade (Fortune Business Insights 2020a).

EU member states needed to adopt a specific regulatory framework to support technologies and practices that will enable the green transition and the shift to a climate-neutral economy (including bioenergy, which accounts for 57.3% of renewable energy sources and 11.4% of total energy consumption in the EU27 today). In November 2016, the European Commission published the "Clean Energy for All Europeans" initiative. As part of this package, the European Commission adopted a legislative proposal to amend the Renewable Energy Directive (RED), which entered into force in December 2018 (revised Renewable Energy Directive 2018/2001/EU [RED II]). The RED II states the need to increase the use of renewable energy sources to achieve the target of 32% by 2030. The European Commission's previous proposal did not include the transport sub-target, which has now been introduced in the final agreement. The EU member states must require fuel suppliers to supply a minimum of 14% of the energy consumed in road and rail transport by 2030 as renewable energy (Renewable Energy-Recast to 2030 [RED II] 2019).

Biofuels are also important for achieving the GHG reduction goals. However, biofuel production usually takes place on arable land previously used for other agricultural crops, such as growing human food or animal feed. Since it is still necessary to produce crops, agricultural land could be expanded to non-arable land, including areas with high carbon stock, such as forests, wetlands, and peatlands. This process is known as indirect land-use change (ILUC). Since such an increase in biofuel production can lead to the release of carbon dioxide stored in trees and soil, ILUC does not contribute to GHG reductions. To address the issue of ILUC, the RED II introduces a novel approach that sets limits on high ILUC-risk biofuels, bioliquids, and biomass fuels. These limits will affect the amount of these fuels that the member states can count towards their national targets when calculating the overall national share of renewables and the share of renewables in transport. Member states will still be able to use (and import) fuels covered by these limits, but they will not be able to include these volumes when calculating the extent to which they have fulfilled their renewable targets. The directive also introduces an exemption from these limits for biofuels, bioliquids, and biomass fuels certified as low ILUC risk. The commission has also adopted an accompanying report on the status of production expansion of relevant food and feed crops worldwide, based on the best available scientific data. The report provides information and criteria for the EU member states to identify high ILUC-risk fuels and certify low ILUC-risk fuels (Renewable Energy-Recast to 2030 (RED II) 2019; Bioenergy Europe 2021). Policy decisions must contribute to the right environment for investment in renewable energy and materials. The lack of policymaking for forest biomass sustainability criteria could lead to re-carbonisation and harm the EU rural economy. Given the progressive nature of RED II and its comprehensive sustainability framework for forest biomass, changes to forest biomass sustainability criteria should be avoided or minimised to ensure that the share of renewable energy continues to grow, which will be necessary for the EU to meet its new target (32% of consumption from renewable energy sources by 2030) (Bioenergy Europe 2021).

Biomass is supplied from various raw materials: wood fuel, forest residues, charcoal, pellets, agricultural crops and residues, municipal and industrial waste, biogas, biofuels, and so on. The supply can generally be classified into three main sectors: forestry, agriculture, and waste (European Commission 2019a).

More than two billion people worldwide still use *wood* as the primary energy source for cooking and/or heating, especially in households in developing countries. This accounts for one-third of global renewable energy consumption and still makes wood the most important single energy source, ensuring around 6% of the total global supply of primary energy (The State of the World's Forests 2020).

Forests cover 31% of the global land area, with approximately half of the forest area relatively intact. More than one-third of the world's forests are primary forests, defined as naturally regenerated forests of native tree species where there are no clearly visible indications of human activity, and the ecological processes are not significantly disturbed. However, more than half of the world's forests are found in only five countries (Brazil, Canada, China, the Russian Federation, and the United States). Today, deforestation and degradation represent the most prominent global issue, and it will not be easy to meet the goal of increasing forest areas by 3% by 2030 included in the UN Strategic Plan for Forests. The data from 1990 to 2020 show that deforestation is decreasing, but insufficiently to stop this trend concerning afforestation (see Figure 4.5) (FAO 2020). These problems, among other things, lead to a significant loss of biodiversity.

Forests provide more than 86 million green jobs and support the livelihoods of even more people. An estimated 880 million people worldwide spend part of their time collecting firewood or producing charcoal, mostly in low-income countries with high poverty rates. Some 252 million people living in forests and savannahs have incomes of less than $1.25 per day. In addition to climate impacts, forests maintain biodiversity, which is essential for the sustainability of food production. It is imperative that we stop deforestation and transform our food systems in terms of producing and consuming food. The demand for food

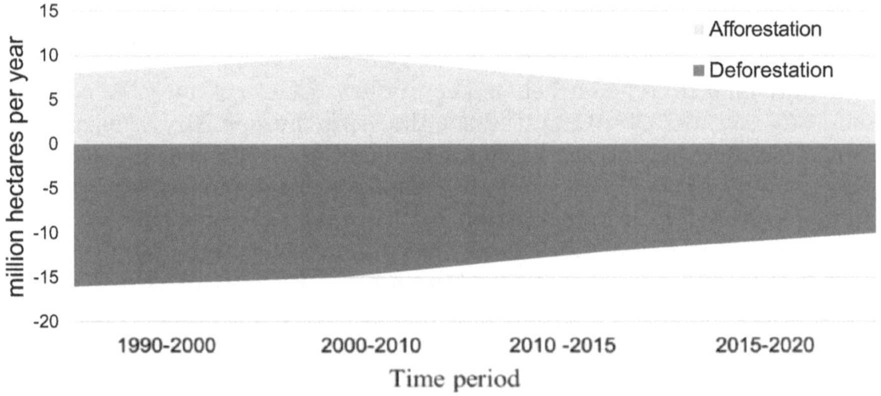

Figure 4.5 Global afforestation and deforestation activities 1990–2020

Source: FAO, Global Forest Resources Assessment 2020

results in inappropriate agricultural practices that turn forests into agricultural land and lead to the loss of forest-related biodiversity. Adopting agroforestry and sustainable production practices, restoring the productivity of degraded agricultural land, adopting healthier diets from sustainable food systems, reducing food loss and waste, and adopting environmental and social responsibility models are some of the actions to be taken urgently. The United Nations (UN) also highlights the importance of restoring ecosystems, as indicated in the UN's latest report for 2021–2030 (The State of the World's Forests 2020).

Forests, other woodlands and trees outside forests, by-products from wood processing, wood recovered after consumption, and processed wood-based fuels are *wood fuel* sources. Wood fuels are a significant forest product. Global production of firewood exceeds the production of industrial roundwood. Firewood and charcoal are often the dominant uses of wood biomass in developing countries and economies in transition. Wood energy is also a vital reserve fuel for emergencies. Societies at any socio-economic level will quickly return to wood-derived energy when faced with economic difficulties, natural disasters, conflict situations, or energy shortages from fossil and/or other sources. Today, the focus has shifted towards wood energy due to climate change and energy security concerns (Wood Energy 2021). Bioenergy is mainly used in homes and to a lesser extent in industry, while liquid biofuels for transport still have limited use (Figure 4.6).

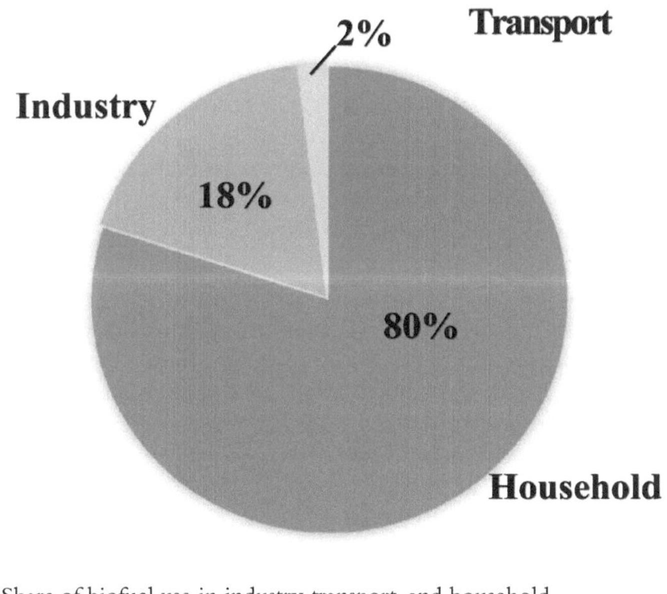

Figure 4.6 Share of biofuel use in industry, transport, and household

Source: Author

Bioenergy market analysis

The growth of the bioenergy market can be attributed to the growing shift to renewable energy sources and other energy sources that are not based on fossil fuels, such as solid biomass, biogas, and others. Factors such as the increasing energy demand, advances in technology, increasing investment in bioenergy, and reductions in the cost of electricity generation from bioenergy plants are expected to boost the market over the analysed periods. However, the growing demand for electric vehicles around the world is also likely to negatively affect the demand for biofuels in the coming years (European Commission 2019b).

Various global market analyses predict trends in the bioenergy market (Research and Markets.com 2021b; Allied Market Research 2018; Research and Markets.com 2021c). According to one analysis (Research and Markets.com 2021b), the bioenergy market is expected to grow by $42.54 billion from 2021 to 2025 (compound annual growth rate [CAGR] of 4.52%), while another analysis predicts it will reach $642.71 billion from 2020 to 2027 (CAGR of 8.0%). A critical factor that will increase the growth of this market relates to the growing commitment of countries around the world to achieving the goals of climate agreements (the 2015 Paris Agreement and the new 2021 Glasgow Agreement). In the effort to control emissions, countries are actively switching to bioenergy sources such as biomass and biofuels to meet their electricity requirements since these sources can supply electricity to all end-use industries. Therefore, expanding participation in the Paris Agreement will be necessary for the bioenergy market (Fortune Business Insights 2020a).

Solid biomass dominated the bioenergy market in 2018 and is expected to continue its dominance during the following period. The Biomass Power Generation Market report is segmented into urban residue, municipal solid waste, agricultural and forest residue, energy crops, woody biomass, and so on (Fortune Business Insights 2020a, 2020e).

Europe dominated the global biomass power generation market during the previous period. The EU member states have included the bioenergy option in their national renewable energy action plans.

In 2018, more than 65% of the biopower generation in Europe came from solid biomass sources, such as wood chips and wood pellets, around 20% came from municipal waste, followed by 15% from biogas. Electricity generated from biomass and waste accounted for approximately 2% of the total electricity generation in the United States. A similar trend has been witnessed in developing countries, such as India and Indonesia. India's biomass power generation capacity has rapidly grown over the last few years as the Indian government focuses on increasing power generation through renewable energy sources. Therefore, with the increasing demand for renewable energy sources across the globe, biomass-based electricity generation is expected to witness a significant growth rate during the analysed period (Research and Markets.com 2021a).

In 2018, the biomass-based installed power capacity increased by 5% over the previous year and has accounted for more than 70% of the bioenergy installed

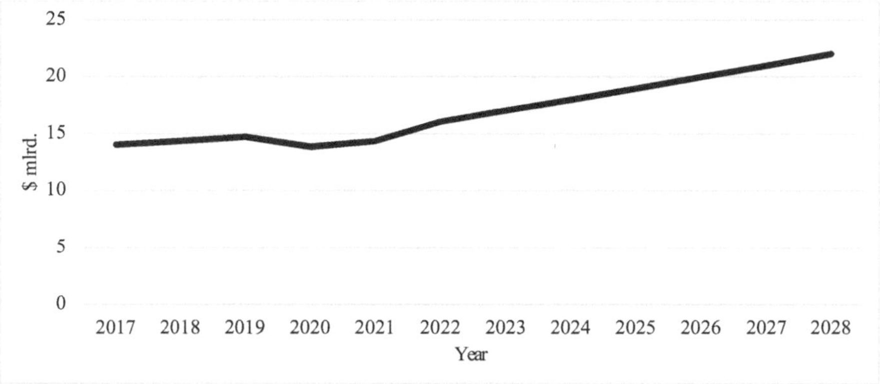

Figure 4.7 Estimation of the biogas market size in the EU for the period 2017–2028. $ mlrd is USD billion.

Source: Author

capacity across the world. Throughout 2018–2023, bioenergy (including liquid biofuels) is projected to account for 30% of the growth in renewable energy production (Reid *et al.* 2020).

As one of the largest producers of liquid biofuels, North America is constantly consolidating its bioenergy market share by significantly increasing bioethanol production.

By strengthening global commitment towards bio-based energy, critical players in this market are exploring new ways to expand their regional and international presence. Likewise, companies are taking advantage of incentives governments offer for renewable energy projects to consolidate their position in the market (Fortune Business Insights 2020a).

Concerning the size of the *biogas* market in the EU, analyses predict further growth of the biogas market in the period from 2017 to 2028 (Figure 4.7) (Fortune Business Insights 2020c).

The global *bioethanol* market has been segmented into North America, Europe, Asia-Pacific, South America, the Middle East, and Africa. The United States is the largest producer, and Brazil is the second-largest bioethanol producer globally. At the same time, China and India are expected to increase the production and use of bioethanol in the coming period. Energy consumption in North America is higher than in South America, so biomethane is used to meet the increasing electricity demand. The leading countries in biomethane production in Europe are Germany, Italy, and France, and until recently, the United Kingdom. Biomethane production has been intensively promoted on a large scale in Southeast Asia. Table 4.3 shows data on the national share of

Table 4.3 National ranking according to biofuel production and share of bioethanol and biodiesel production

Biofuel Production			
	Bioethanol, %		*Biodiesel, %*
United States	50	EU	36
Brazil	24	United States	19
China	8	Brazil	12
EU	4	Indonesia	10
India	2	Argentina	7
Canada	1.6	Thailand	4
Thailand	1.5	China	3
Argentina	1	Colombia	1.5
Colombia	0.4	Canada	1.4
Paraguay	0.3	India	0.5
Indonesia	0.2	Paraguay	0.03

biofuel production, ranked from the highest to the lowest share of bioethanol and biodiesel production (Fortune Business Insights 2020b, 2020d).

Replacing fossil fuels with biofuels is necessary for decarbonising the transport sector (truck, ocean, or air transportation). Most biofuels are currently consumed through blending at low percentages with fossil fuels (typically less than 10%). Biofuel demand grew 5% per year on average between 2010 and 2019, while the Net Zero Emissions by 2050 Scenario requires a much higher average growth of 14% per year to 2030. Biofuel demand and production are expanding globally, but not at a pace consistent with the Net Zero Scenario, which is why policies should include rigorous sustainability criteria to contribute to social, economic, and environmental progress, including reducing GHG emissions and finding ways to make higher shares of biofuels to replace gasoline or diesel and contribute to greater commercialisation of advanced biofuels (IEA Tracking Clean Energy Progress 2021).

Innovative technologies in the bioenergy sector

Innovations in the bioenergy sector refer to innovative and sustainable technologies that convert distinct types of waste and biomass into biofuels. In addition to waste, new chemicals are being developed to contribute to novel liquid biofuels. New innovative solutions and technologies in the bioenergy sector are pushing the boundaries of conventional science, whereby renewable biofuels could provide a cost-effective alternative to fossil fuels to reduce harmful GHG emissions and consequently mitigate climate change. Investment in research and innovation in the bioenergy sector is estimated to increase added value tenfold.

However, despite significant advances in innovative technologies that use biofuels to produce bioenergy, the environmental impacts of the production of liquid biofuels for transport, policy costs for their promotion, and possible adverse effects have been severely criticised. Although the production of biofuels (ethanol and biodiesel) is growing, it is assumed that their contribution to total production-consumption will be limited in the coming decades. In addition, despite the significant impacts of the increased biofuel production on global agricultural markets, food safety, and the environment, the use of biofuels continues to cause a great deal of controversy (Trzmielak and Kochańska 2021, p. 413; Department for Business, Energy and Industrial Strategy 2019).

Given the need to reduce waste and minimise the use of fossil fuels (i.e., shift to technologies that use renewables instead of those that use fossil fuels to prevent further climate change), preserve the environment, while ensuring economically viable alternatives, following are some examples of innovative technologies for bioenergy production (Office of Energy Efficiency & Renewable Energy 2016; Mordor Intelligence 2020a; Ductor – Our projects 2021; Brightmark 2021 – The Yellowjacket project in renewable natural gas; Greenbelt Resources Corporation 2019; Clariant 2016; European Biogas Association, 2018; Carbon Recycling International n.d.; FuelCellsWorks 2020):

- Commercial airline Virgin Atlantic (United Kingdom) announced that it would use renewable jet fuel produced by LanzaTech. This process converts carbon-rich industrial waste gases (such as carbon monoxide from steel production) into aircraft fuel. LanzaTech (United States) and the Pacific Northwest National Laboratory (United States) developed this process to provide a sustainable source of renewable jet fuel to reduce GHG emissions in the transport sector and provide an innovative solution for industrial waste management.
- Researchers at Sandia National Laboratories (New Mexico, United States) are developing "Algal Turf Scrubbers" – long, thread-like algae that could be installed near rainwater or wastewater runoff sites to absorb excess phosphorus and nitrogen from agricultural sites. Excess phosphorus and nitrogen can harm aquatic ecosystems, and Algal Turf Scrubbers can mitigate this problem. The algae can then be converted into biofuel.
- The Pacific Northwest National Laboratory (United States) and industry partners are working to develop technology to produce biofuel from wet wastes at water treatment plants as raw material. Each year the United States treats approximately 42 billion litres of municipal wastewater, generating about seven million tons of dry sewage sludge. Managing and disposing of the resulting sludge is very costly and accounts for almost 45%–65% of a treatment plant's total operating expenses. Converting this sludge into renewable fuels could provide an economically viable alternative and transform the existing water treatment plants into a new generation of water resource recovery facilities.
- Emerging technologies for converting waste into energy, such as Dendro Liquid Energy (Germany), a recent German innovation in the biological treatment of waste, present high potential in the bioenergy field, close to "zero-waste" technology.

- Ductor (Finland, Switzerland) announced it would build three new biofertilizer-biogas facilities in Poland's Zachodniopomorskie region to use poultry waste. The facilities will produce renewable electricity and organic nitrogen fertiliser, intensifying Poland's shift towards a circular economy.
- Brightmark (United States) has expanded its dairy biogas project within the Yellowjacket project. In partnership with six farms, the company will daily extract methane generated from dairy waste and produce renewable biogas.
- Greenbelt Resources (United States) adds other milk derivative wastes to the dairy processing waste to prompt the so-called New Age of renewable energy.
- Clariant (Romania), a world leader in speciality chemicals, has built an innovative bioethanol plant.
- Spanish car company SEAT has announced a project to turn wastewater into biomethane.
- The George Olah Renewable Methanol Plant (Iceland) was commissioned in 2011 to use geothermal energy to produce renewable methanol that is miscible in gasoline. It is also used in biodiesel production and other methanol-based processes. In addition, plans to open a factory in China have been announced.
- A team of researchers from the Ulsan National Institute of Science and Technology (North Korea) has developed a new system for efficient hydrogen production. The new fuel is obtained from biomass (lignin).

The bio-based chemistry sector

Green chemistry has been evolving as a new discipline since the 1980s, focusing on using less hazardous chemical syntheses, designing environmentally benign chemicals, using safer solvents, and optimising processes with improved energy efficiency. The transition from fossil to renewable raw materials has begun by broadening chemical catalysts to reduce emissions and improve process efficiency. Today, the platform of knowledge gained through the evolution of green chemistry is integrated into bio-based technologies. The low oil price has previously represented an obstacle to the production of bio-based chemicals, and chemical manufacturers have consequently based their bio-based production on chemicals with specific properties and complexity of structure to justify production costs (Lange *et al.* 2021; IFP Energies Nouvelles 2021).

Bio-based chemicals, which are defined as chemical products that are wholly or partly derived from materials of biological origin (such as plants, algae, crops, trees, marine organisms, and biological waste), will play a significant role in the future. Given their expected limited environmental footprint compared to their traditional counterparts, bio-based chemicals have recently emerged on markets as environmentally friendly alternatives to standard chemicals (EU Science Hub 2019).

The development of bio-based chemicals and polymers leads to new requirements for raw materials, the development of new technologies, and new economic opportunities. At the same time, wider application faces challenges related to production costs, security of access to resources, development of the supply sector, use of renewable sources in production, and recyclable product design. On the one hand, the development of bio-based chemicals is constrained by the lower price of fossil fuels compared to the cost of renewables, trade wars, a lack of policy and legislative coherence, and insufficient or no sanctions for GHG emissions for those who continue to produce fossil fuels. On the other hand, it points to the need to reduce the chemical sector's dependence on fossil fuels to contribute to the reduction of GHGs and stimulate the circular economy to achieve sustainability and human health (Philip *et al.* 2013).

Despite successful applications in several industries (cosmetics, solvents, dyes, and other standard products), bio-based chemicals still have limited market share and investor interest. However, the future investment in bio-based production capacities is currently uncertain and depends on a complex combination of factors such as future oil prices, consumer behaviour, and technological development. Therefore, predictions regarding the future markets are intended as indicative extrapolations based on current trends. The production of chemicals and polymers from renewables is essential for environmental sustainability, but it is also an important part of the transition to a circular economy. Figure 4.8 shows product categories (I–X) covering the field of bio-based chemistry, as well as their predicted increase in production in 2025 (Spekreijse *et al.* 2019).

In addition to the production of bioenergy or biofuels based on agricultural products or forest, marine, or solid waste, biorefineries offer various opportunities for the development of bio-based products (chemicals and polymers) and materials (fibres, starch derivatives) that are produced as products or as by-products depending on technological processes (fermentation, thermochemical conversion, pyrolysis, degassing, acid and ester hydrolysis, chemical

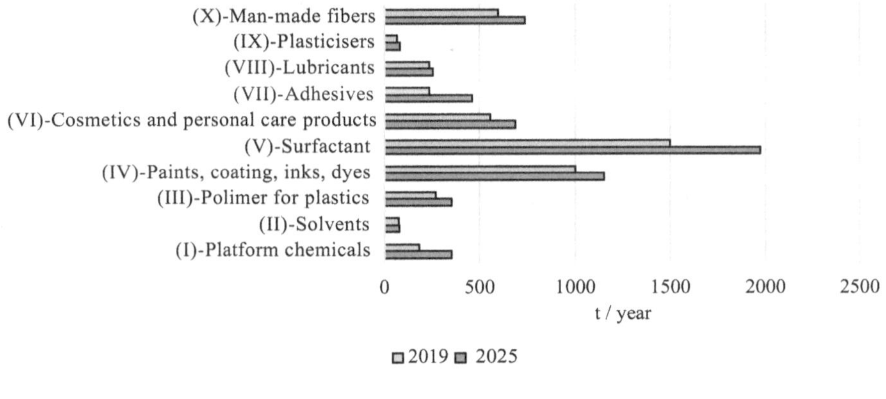

Figure 4.8 Bio-based production in 2019 and the predicted bio-based output in 2025 for designated product categories

Source: Insights into the European market for bio-based chemicals, Spekreijse *et al.* 2019

conversion, anaerobic digestion, transesterification, etc.). For example, adipic acid, 3-hydroxypropionic acid/aldehyde, isoprene/farnesene, glutamic acid, aspartic acid, and 1,4-butanediol are innovative chemicals obtained in the fermentation process. The processes of obtaining chemicals from lignocellulose biorefineries and using biogas for microbial fermentation, which would produce alcohols, acids, and acid derivatives, are being intensively developed. However, the cost of bio-based solvents often still exceeds the cost of solvents obtained from petrochemical sources (i.e., they have a higher starting price), and new production processes need to be optimised.

R&D of new technologies and legal regulations support the *fermentation products* market growth. Alcohols, amino acids, polymers, industrial enzymes, and vitamins are essential components used in fermentation to produce bio-based products. In the food and beverage industry, these products help store food and extend the product's shelf-life. In addition, the increasing use of fermentation products as raw materials in the clothing, leather, plastics, chemical, and rubber industries and the increase in demand for antibiotics and steroids contribute to the development of the fermentation products market.

There has recently been an increase in the use of *bioplastics* due to the growing awareness of consumers about the harmfulness of conventional products (plastics obtained using fossil fuels) to human health. In addition, landfills destined for plastics obtained from the petrochemical industry have a detrimental effect on flora and fauna (decomposition takes several hundred years), making the use of biodegradable plastics that can be re-absorbed into the ecosystem necessary for the preservation of the environment.

Biopolymers (macromolecules derived from plants, trees, bacteria, algae, or other natural sources) can be produced through several different processes, such as the genetic modification of plants, starch conversion, or microbial conversion. There are three categories of biopolymers depending on the origin of raw materials and the production. These include biopolymers derived from polysaccharides (starch, chitin), biopolymers produced from classical chemical synthesis using renewables, and biopolymers produced by microorganisms or genetically modified bacteria. For example, the most commercially available polylactic acid (PLA) biopolymer is produced from lactic acid in the dextrose fermentation process. The development of biopolymers is intensive, and innovative biopolymers have already been used in electronics, construction, paper, packaging, cement, automotive, and other industries.

The "Environmental Impact Assessments of Innovative Bio-based Products" (EU, JRC) study analysed three main commercialised biopolymers: (1) bio-based polyethylene terephthalate ("Beverage Bottles" using a bio-based substitute for monoethylene glycol); (2) PLA ("Single-use cups", "Single-use Cutlery", "Packaging films"); and (3) starch plastics ("Clips", "Mulch films" for mulching soil, and "Carrier bags"). The data are gathered from industry based on the actual supply chain and include the currently used biomass. This study indicates that exceptionally valuable properties and efficient use characterise these biopolymers compared to polymers produced from classical chemical processes. In addition to being effective in terms of application and sustainability, they also have a lower environmental impact (reduced GHG emissions) because of the use of renewables (biomass) in their production.

Lignin, a complex organic polymer and the second-largest carbon source, is a by-product of the pulp and paper industry. The development of new technologies increases the availability of lignin for applications outside of the pulp and paper industry. It is estimated that 50–70 million tons of lignin are produced annually, but its consumption is exclusively related to the said industry. Lignin is widely used in the construction industry (additives for concrete, binders, bitumen, adhesives) and the production of insecticides and pesticides, animal feed, and products for personal use. In addition, the use of lignin represents a significant opportunity to improve lignocellulose biorefineries, while new research and innovations enable higher jet fuel production.

As non-renewable resources are increasingly depleted, and countries worldwide impose stricter environmental laws that affect human health, *oleochemicals*, products obtained from natural oils and fats, represent environmentally friendly products and sustainable alternatives to conventional products. Due to their availability, oleochemicals can be used in a broad range of industries and products, such as metalworking lubricants, additives for plastics, rubber and paper production, automotive lubricants, greases and fuel additives, and personal care products like soaps, lotions, cosmetics, and deodorants. Their growing demand results from public awareness of the need for environmental conservation, acceptable production costs, and rising oil prices (Golden and Handfield 2014; IEA Bioenergy 2020; Univar Solutions n.d.).

Bio-based chemistry market analysis

Bio-based chemicals' analyses are conducted through numerous studies. One such study is the BREW Project that analyses the market potential of renewable bulk chemicals in biotechnology. According to the study, the production of organic chemicals could, under favourable conditions, grow to 113 million tons by 2050, which would represent 38% of the total organic chemical production. Under unfavourable conditions, the production could increase to 26 million tons (i.e., 17.5% of organic chemical production). The economic impact, functionality, ease of use, and cost of producing such bio-based chemicals are assessed in relation to petrochemicals from an economic perspective. It takes 30 years to implement bio-based chemicals (to reach the full economic potential) if the value of the bio-based product is not lower than the depreciated production costs of the petrochemical counterpart. If the value of bio-based chemical products is lower than the value of depreciated production costs of the petrochemical counterpart, that period is reduced to 10 years (Patel *et al.* 2016).

Global production of bio-based chemicals and polymers is estimated at around 90 million tons, while global petrochemical production of chemicals and polymers is estimated at approximately 330 million tons. Furthermore, the production of bio-based chemicals could generate more than $10 billion in revenue for the global chemical industry. However, there are certain restrictions, such as market conditions, trade agreements, carbon prices, and so on (Spekreijse *et al.* 2019). In addition to these obstacles, the entire bio-based chemicals and polymers market is currently affected by the Covid-19 pandemic.

Limitations in the production of *biopolymers and bioplastic* used in various industries such as packaging, consumer goods, automotive industry, transportation, textiles, agriculture, and so on, has led to a sharp decline in demand for some types of packaging. On the other hand, the demand for essential packaging such as e-commerce shipment has experienced tremendous growth. Due to the global supply chain disruptions, 76% of the businesses had to reduce revenue targets by an average of 23%. Consumer awareness regarding sustainable plastic solutions and comprehensive efforts to eliminate the use of conventional plastics contribute to the growth of the biodegradable bioplastics market (Markets and Markets 2021a).

Increasing consumer awareness of organic products and environmental issues and the growth of the bioethanol industry are driving the growth and development of the demand for *fermentation* products. Thanks to the growing demand for fermentation products in the food and beverage industry, the Asia-Pacific region dominates the fermentation products market. In developing countries, such as India, China, and Brazil, the demand for fermentation products, technical progress, and growth in the number of applications of these products is projected to provide opportunities for business expansion (Industry ARC 2021).

The fact that *oleochemicals* (fatty acids, fatty alcohols, glycerine, etc.), which are widely used in the production of paints, detergents, soaps, lubricants, copolymers, and so on, can replace many petroleum-based products, are environmentally friendly, and can be manufactured from renewables is expected to further the demand for these sustainable and biodegradable products. New, innovative products are also likely to contribute to the growth of the oleochemical market. Key players in the global oleochemical market are present mainly in Indonesia, Malaysia, and the Philippines (Markets and Markets 2015).

Most *lignin*, a by-product in the production of paper and pulp, appears in the form of ligno-sulphonates and kraft lignin. The annual lignin production is estimated at around 50–70 million tons. Although it is mainly used as an energy source in paper and pulp mills, the development of new innovative technologies increases the application of lignin to other areas such as the production of chemicals, resins, and polymers. One of the examples refers to using lignin-based renewable functional fillers in rubbers as a sustainable alternative to carbon black and silica. Lignin also represents a significant opportunity to improve lignocellulose biorefineries. North America dominates the global lignin market due to the growing demand for concrete and cement. Europe represents the second-largest lignin market due to the increasing demand for lignin in the end-use industry (Market Research Engine 2021; GlobeNewsWire 2021; Global Market Insights 2020; CISION PR Newswire 2021a).

Although new technologies in the bio-based chemistry sector are still emerging, novel bio-based products are helping increase the demand for more sustainable end-use goods. Efforts are also being made to use 100% bio-polyethylene terephthalate packaging and 100% bio-nylon textile fibres. The lack of raw materials (organic chemicals) also contributes to the shift towards new raw materials and production processes. However, the market will continue to depend on crucial economic and political factors and the price of oil.

Global market growth forecasts and estimates of CAGRs (%) for bioplastics and biopolymers, fermentation products, oleochemicals, and lignin for the analysed periods are shown in Table 4.4. However, these forecasts depend on political, economic, and environmental factors, as mentioned earlier (CISION PR Newswire 2021b).

The projected significant global market growth referring to bio-based product categories I–X in the EU (Spekreijse *et al.* 2019) is shown with CAGRs over the five-year period from 2020 to 2025 (Figure 4.9).

Table 4.4 Market growth forecasts and estimates of CAGR (%) for bioplastics and biopolymers, fermentation products, oleochemicals, and lignin for the analysed periods

Product	*Analysed period*	*Global market assessment*	*CAGR*
Biopolymers and bioplastics	2021	$10.7 billion	22.7%
	2026	$29.7 billion	
Fermentation products	2016	$149.5 million	4.7%
	2023	$205.5 million	
Oleochemicals	2020	$25.8 million	6.5%
	2027	$39.8 million	
Lignin	2020	$770 million	5.5%
	2027	$1.2 billion	

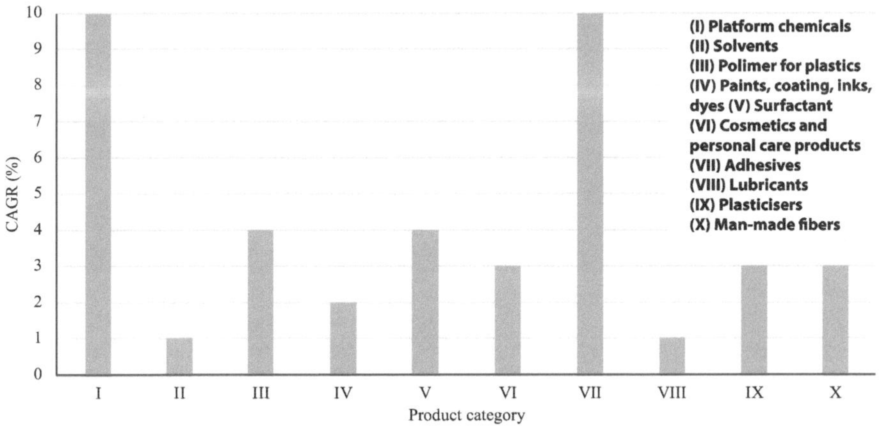

Figure 4.9 CAGR for product categories I–X over the 2020–2025 forecast period in the EU

Source: Insights into the European market for bio-based chemicals, Spekreijse *et al.* 2019

Innovative technologies in the bio-based chemical sector

Innovative technologies in the bio-based chemical sector will enable the construction of a future biomass-based industry because, in addition to biomass fuels, they can help produce new products (chemicals, polymers) and new materials as innovative products. The development of new products also brings about the development of new business models. Since this is an emerging industry, product and process innovations occur intensely, there are no dominant designs or technologies, and competitive patterns have not yet been established (Bomtempo and Alves 2014).

By 2050, the potential substitute for large quantities of bio-based chemicals has been estimated, under favourable conditions, at around 38% of the organic chemicals market, as mentioned earlier.

Given the potential to develop new products in the bio-based chemical sector, especially in biorefinery processes, following are some innovative technologies to produce biochemicals and bioproducts (Smith 2019; Adler 2021; AD Bioplastics 2020; Valmet 2021; Bio Explorer 2018; ZELCOR 2021; NEWFERT 2018; Acme-Hardesty 2020; Carbios 2020; Basilisk Self-Healing Concrete 2021):

- Afyren (France) commercialises a technology that uses bacteria to produce industrial chemicals from recycled plant waste instead of fossil fuels. Sugars from waste by-products of agriculture, food, and other industries are fermented into industrial chemicals using bacteria. The chemicals Afyren produces include acetic and butyric acid. Many of the chemicals produced by Afyren are useful in the food, agricultural, and chemical industries, the same industries from which the company sources its waste sugars.
- The Carbon4PUR project has solved two major challenges, one of which relates to the conversion of gases in steel plants into higher-value products for market-oriented consumer products. In contrast, the other relates to a new process based on direct chemical conversion of gas mixtures produced in steel plants. Both challenges are characterised by significant savings and a reduced carbon footprint.
- ADBioplastics is a Spanish start-up that develops bioplastics. It produces premium-grade PLA from corn, sugar cane, or sugar beet. The patented BlockPLA technology improves conventional PLA qualities (improved properties, biodegradability, and compostability). The resulting product is amenable to thermoforming, injection, and injection blowing and is used in various sectors.
- Woodoo is a French start-up that uses bioengineering to produce augmented and translucent wood. The technology removes air and lignin from the wood while maintaining its structural integrity and reinforces the wood with a bio-based polymer. It enables the use of otherwise unused, low-grade wood in construction materials. The extracted high-quality lignin is a bio-based polymer that can produce speciality chemicals.

- Puraffinity is a British start-up that develops advanced materials for environmental applications. It combines advances in bio-based polymers and supramolecular chemistry to produce new adsorbents. Customised granular media, a biodegradable polymer, removes most suspended solids, such as per- and polyfluoroalkyl substances, which are non-degradable and harmful to human health and cannot be removed with existing filtration techniques, from water samples. Customised granular media is also regenerable, easy to use, and has a low carbon footprint.
- LignoBoost is a technology that allows lignin to be extracted from the Kraft pulping process. LignoBoost is proven and reliable and has been developed with several industry partners. There are four LignoBoost plants in operation today. The LignoBoost plants in Finland and the United States were built by Valmet for commercial lignin production and have a combined design capacity of 75,000 tonnes of lignin per year. Valmet also recently built a smaller plant in Brazil.
- Researchers at the University of South Carolina (United States) have devised an innovation that uses grafted block copolymers, star copolymers, and multi-segmented block copolymers to prepare microphase-separated films. They can be further used to prepare templates for nanolithography. This innovation combines both high-resolution and high-quality films needed for lithography.
- The *Zelcor* project team (Finland, France, Italy, the Netherlands, Germany, Switzerland, England) combined chemical and enzymatic catalysis with insects-based biological conversion, demonstrating the feasibility of converting recalcitrant lignocellulose biomass by-products into high-added-value products.
- The *Newfert* project (Spain, Germany, France) developed new chemical and bio-electrochemical technologies to recover nutrients from ashes of different origins and livestock effluents, which are then used to produce advanced fertilisers.
- Since the autumn of 2011, Sardinia has been working on bio-based chemistry products made from a local plant species known as the cardoon. The cardoon is a thistle-like plant closely related to the artichoke and actively grown in Porto Torres in northern Sardinia. This herbaceous perennial species originates from the Mediterranean and produces seeds with excellent oil, which is the perfect building block for numerous bio-based products such as biochemicals, bio-intermediates, bases for lubricants, and bio-additives for rubbers. Currently, the most popular use for this thistle oil is in the production of bioplastics. Reactors convert vegetable oil into acids that are raw materials for bioplastics, biochemicals, and other products. The Acme-Hardesty plant was built in an abandoned petrochemical plant. Today, they produce bioplastics. They also produce several bio-lubricants with thistle oil, including one specifically for ships. This ship bio-lubricant is non-polluting and dissolves in the sea, making it environmentally friendly. Europe alone produces three million tons

of lubricating oil made from petrol that ends up in the ocean each year. Instead of these ecologically damaging pollutants, bio-lubricants decompose naturally, resulting in minimal impact on aquatic ecosystems.

- In a world where plastic recycling and zero waste are becoming increasingly important, Carbios (France) is one of the most advanced biotechnology companies developing ways to produce and recycle environmentally friendly plastics. Carbios has produced 100% enzymatically recycled plastic bottles. In June 2020, the company announced the construction of its industrial pilot plant, which will allow it to commercialise polyethylene terephthalate packaging recycling technology. The biotechnology company expects to begin the first phase of operations in the first quarter of 2021.
- Green Basilisk (Netherlands) produces self-healing concrete using limestone-producing bacteria. This material has a longer service life than ordinary concrete and requires less maintenance. In addition, the company has begun demonstrating evidence of the concept of new technology.

The biochemistry and biopharmaceutical sector

Biochemistry, or the chemistry of living systems, encompasses multiple fields such as biology, molecular biology, molecular genetics, microbiology, and organic and inorganic chemistry. Experts in the field of biochemistry (biochemists) investigate the structure, composition, chemical processes, and chemical reactions in living organisms; analyse chemical reactions in cells and tissues of living beings; study the genetic structure; and investigate the effects of food, drugs, and other substances on living tissues.

Certain biochemistry studies have highlighted some biochemistry branches, such as clinical and nutritional biochemistry, medicine and pharmaceutical biochemistry, structural and molecular biochemistry, and protein and analytical biochemistry. Research is focused on the study of physico-chemical and biological processes of enzymes, lipids, proteins, nucleic acids, chromosomes, and so on, using cutting-edge instrumental techniques (electron microscope, laser, and state-of-the-art laboratory instruments) and relying on the development of new chemical and diagnostic tools, platforms, models, and new technologies in the field of biochemistry (Schepartz 2018; Alhyari 2018; Glitsoe *et al.* 2015; Moghe *et al.* 2017).

Innovations in biochemistry are necessary for the application of better solutions that meet new requirements or existing market needs, that is, more efficient solutions for processes, products, services, or innovative ideas in biochemistry but also the possibilities of their application in the production of functional, structural materials, and other areas, such as agricultural and waste industries. One such innovation refers to bio-based polymer nanofibers that can be applied in biomedicine, which is why many companies and start-ups are investigating their properties such as biocompatibility, biodegradability, antibacterial activity, and low immunogenicity. Biochemistry research has increased due to new methodologies, molecular modelling techniques, and advanced tools and technologies (Dou 2020).

Biopharmacy covers the study of the relationship between physical and chemical properties, dosage and form of product administration, and its activity in a living body. Research, innovation, and clinical trials in humans and animals are crucial to drug development and marketing. There are several groups of drugs, such as innovative (originator) chemically derived drugs, generic drugs that are copies of innovative drugs, biologics, biotech drugs (including a wide range of products such as vaccines, therapeutic proteins, blood and blood components, tissues, etc.), and biosimilar drugs that are versions of biologic products. In contrast to chemically synthesised drugs, which have a well-defined structure and can be thoroughly verified, biologics are derived from living material (human, animal, microorganism, or plant) and are vastly larger and more complex in structure. Biologics are revolutionising the treatment of cancer and autoimmune disorders and are vital to the industry's future (SelectUsa 2021).

The development of genetic engineering in biotechnology, medicine, and biology in the 1970s made it possible to make changes in the organisms' genetic material (DNA). Research activities and the development of oligonucleotide synthesis have enabled the molecular-level design and modification of biological systems, which has led to the expansion of the global oligonucleotide synthesis market. The increasing prevalence of chronic and life-threatening diseases has prompted pharmaceutical and research companies to focus on R&D activities to develop effective therapies (Journal of Biomedical Sciences 2019).

Biopharmaceuticals are medical drugs produced using biotechnology. They are proteins (including antibodies) and nucleic acids (DNA, RNA, or antisense oligonucleotides) used for therapeutic or in vivo diagnostic purposes. They are produced by means other than direct extraction from a native biological source. Most biopharmaceutical products are drugs that are derived from biological sources. Biopharmaceuticals (or their components) are isolated from biological sources: humans, animals, plants, fungi, or microorganisms. They can be used in medicinal products for human and veterinary use (BioProcess Online 2006).

Biopharmaceuticals (biological), which consist of sugars, proteins, nucleic acids, living cells, or tissues, are extracted or semi-synthesised from biological sources such as humans, animals, or microorganisms. Unlike traditional drugs synthesised from chemical processes, most biopharmaceuticals are derived from biological processes, including the extraction from living systems or the production by recombinant DNA technologies or genetic engineering. Genetically modified organisms (plants, animals, or microorganisms) are potentially used to produce biopharmaceuticals (Chen and Yeh 2018). Table 4.5 shows the sources of biopharmaceuticals, their characteristics, and therapeutic indications.

Emerging sectors include precision medicine and regenerative medicine, representing completely new ways of treating disease. Precision medicine entails treatment approaches that consider individual variability in genes, environment, and lifestyle. Regenerative medicine includes rapidly evolving technologies such as cellular and gene therapies and biomaterials made from tissues

Table 4.5 Primary sources of biopharmaceuticals, characteristics, examples, and therapeutic indications

Biopharmaceutical Sources			
	Extraction from Living Sources		*Produced by Recombinant DNA*
Characteristics	Conventional biopharmaceuticals are extracted from animals or humans particularly. Some biopharmaceuticals were extracted from animals but are now produced by biotechnology. For example, the therapeutic insulin previously extracted from porcine pancreatic islets is now produced by recombinant DNA technologies in the yeast (*Saccharomyces cerevisiae*) or *Escherichia coli*.	Characteristics	Type I: Substances that are almost identical to key body proteins. Type II: Monoclonal antibodies similar to antibodies produced by the human immune system against microbes. Type III: Receptor constructs (fusion proteins) that are usually based on a naturally occurring receptor linked to the immunoglobulin frame.
Examples	Blood and blood components, organs and tissue transplants, stem cells, antibodies for passive immunisation, faecal microbiota, human breast milk, human reproductive cells	Examples	Blood factors, tissue plasminogen activators, hormones, growth factors, interferons, interleukin-based products, vaccines, monoclonal antibodies, tumour necrosis factors, therapeutic enzymes
Therapeutic indications	Diabetes mellitus Thrombocytopenia Venous thrombosis Growth hormone deficiency Acromegaly Immunisation against diphtheria, tetanus, pertussis, polio, hepatitis B	Therapeutic indications	Haemophilia Severe sepsis Infertility Osteoporosis Hypocalcemia Anaemia Hepatitis C Multiple sclerosis

that repair or replace cells, tissues, or organs. These technologies, which can cure diseases and not just slow their progression or manage symptoms, represent the next major innovation in healthcare (BioProcess Online 2006).

Biopharmaceuticals have the potential to play a vital role in increasing global health to achieve the 2030 Sustainable Development Goals. Herbal drug production has enormous potential and could become an essential system for various new biopharmaceutical products.

Modern biotechnology has led to a resurgence of interest in producing new therapeutic agents from plants. As early as 2001, it had been predicted that the demand for biopharmaceutical products would increase systematically and that it would be wise to ensure availability in significantly larger quantities and at affordable prices (Goldstein and Thomas 2004).

The process of obtaining edible vaccines is based on oral administration of the vaccine, through consumption of the part of the plant that contains the vaccine (tubers, fruits, leaves, etc.) such as potatoes, tomatoes, lettuce, carrots, bananas, and algae. In addition, many studies have shown that genes encoding bacterial and viral antigens are correctly expressed in edible tissues to form immunogenic proteins and stimulate the production of specific antibodies. Thus, technology has always been limited by low levels of transgene expression. However, recent advances in plant transformation have shown the efficacy, stability, and immunogenicity of orally administered antigens (Miranda *et al.* 2020).

Biochemistry and biopharmaceutical market analysis

Advanced instrumental (analytical) techniques, including instruments and biochemical reagents used in research institutes, biotech companies, diagnostic centres, hospitals, pharmaceutical companies, and so on, are essential for conducting biochemical research and stimulating innovation in biochemistry. Therefore, when it comes to the analysis of the biochemistry market, it can be monitored through the biochemistry analysers market, the mass spectrometry market, as well as the biochemical reagents market.

Market analysis assessments were conducted in light of the Covid-19 pandemic and the economic crisis it caused. The global market for biochemistry analysers was estimated at $3.5 billion in 2020 and is projected to reach $4.7 billion by 2026. The United States represents the largest market for biochemistry analysers, followed by China, Japan, and Canada. Germany is the largest and fastest-growing market for biochemistry analysers within the EU. The global mass spectrometry market size was valued at $5,885.28 million in 2020 and is projected to reach $12,697.21 million by 2030. North America represents the largest market for biochemical reagents, while the Asia-Pacific region shall be the fastest-growing market for biochemical reagents in the analysed five-year period. Projections of CAGRs for the biochemistry analysers market, the mass spectrometry market, and the biochemical reagents market are shown in Figure 4.10 (PRNewswire 2021; Allied Market Research 2021; Billings 2021; Mordor Intelligence 2020b).

The global biopharmaceuticals market accounted for $186,470 million in 2017 and is projected to reach $526,008 million by 2025 (CAGR of 13.8% from 2018 to 2025).

The global biopharmaceuticals market is driven by numerous factors, such as the increase in the elderly population, surge in prevalence of chronic diseases such as cancer and diabetes, and global adoption of biopharmaceuticals. Furthermore, the increase in strategic cooperation among biopharmaceutical companies is expected to supplement the growth of the biopharmaceuticals

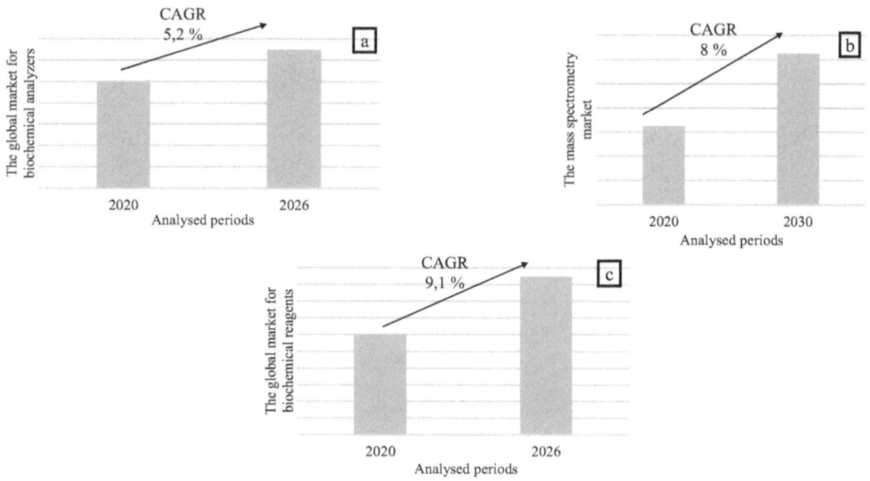

Figure 4.10 Projections of compound annual growth rates (CAGR) for (a) the biochemistry analysers market, (b) the mass spectrometry market, and (c) the biochemical reagents market in the analysed periods

Source: Author

industry. The United States is the largest biopharmaceuticals market, accounting for about a third of the global market and leading in biopharmaceutical R&D. US companies conduct more than half of the world's R&D in pharmaceuticals ($75 billion) and hold the intellectual property rights to most new medicines. However, high costs associated with drug development and risky manufacturing are expected to restrain the growth of the global biopharmaceuticals market. Conversely, emerging economies, such as India and China, are expected to expand strongly into the global biopharmaceuticals market during the forecast period. The global biopharmaceuticals market is segmented by type, application, and region. Based on type, the market is divided into monoclonal antibody, interferon, insulin, growth and coagulation factor, erythropoietin, vaccine, hormone, and others (Figure 4.11). Regarding the application, the market is categorised into oncology, blood disorder, metabolic disease, infectious disease, cardiovascular disease, neurological disease, immunology, and others. Region-wise, the global biopharmaceutical market is divided into North America, Europe, Asia-Pacific, Latin America, the Middle East, and Africa (Market Data Forecast 2021-Report).

Currently, 65% of approved drugs come from biopharmaceutical companies, which is expected to increase to 85% within ten years. The biopharmaceutical industry, particularly innovative SME, has taken centre stage in the fight against the Covid-19 virus. Small biopharmaceutical companies possess several market

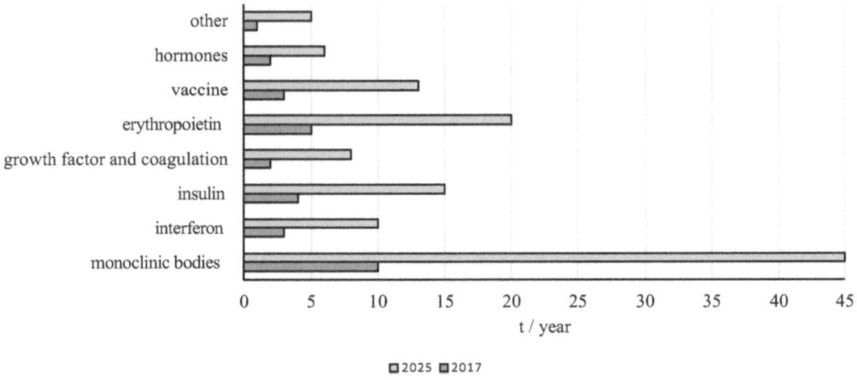

Figure 4.11 Projection of the global biopharmaceuticals market by type of biopharmaceuticals for 2025 compared to 2017

Source: Author

advantages; they are flexible, adaptable, and embrace novel approaches in medicine. In 2017, about 65% of the drugs approved on the market were based on biotechnology, according to United Nations (UN) statistics, and this trend is expected to accelerate in the coming years. The massive demand for biopharmaceuticals will accelerate research and innovation, fuelled by the Covid-19 pandemic. In addition, the ability of biopharmaceuticals to address previously untreatable diseases will spur the introduction of innovative drugs into the market, strengthening the biopharmaceutical market in the next decade. Europe is a global power at the research level with world-class research institutions, medical centres, and hospitals that provide a solid basis for sourcing and developing scientific and clinical innovations. However, the pandemic has also shown what is possible when companies and scientists work together. In the coming years, if innovation continues at the scale seen during the Covid-19 pandemic, or if production is to be made more efficient and cost-effective, the only way forward is to expand partnerships between the countries (Bouzidi 2021).

The most significant innovation emerging from Europe during the pandemic caused by the Sars-CoV-2 virus was the AstraZeneca/Oxford vaccine (United Kingdom). The size of the global antiviral drug market is expected to reach $77.07 billion in 2028 and register a CAGR of 4.6% over the forecast period (2018–2028) (BioSpace 2021).

Intensive research activities in the pharmaceutical and biotechnology sectors have led to a global expansion of the oligonucleotide synthesis market (nucleic acid polymers with the potential to cure or treat a wide range of diseases). The global oligonucleotide synthesis market is projected to reach $14.1 billion by

2026 from $6.3 billion in 2021, at a CAGR of 17.6% over the forecast period (Markets and Markets 2021b).

Innovative technologies in the biochemistry and biopharmaceutical sectors

Since biochemistry has become the foundation for understanding all biological processes and seeks to explain the complex chemical reactions in various life forms, it provides the basis for practical advances in medicine, veterinary medicine, agriculture, and other fields.

Given the wide-reaching areas of research, which, in addition to chemistry and biology, include the fields of medical sciences, numerous innovations in biochemistry provide explanations for the causes of and solutions to many diseases in humans, animals, and plants. Advances in these areas have created links between technology, chemical engineering, and biochemistry. The following examples of *innovations in biochemistry* highlight the solutions that meet new market requirements or needs (Wei *et al.* 2017; Celik and Kilinc 2016; Balskus and Jacobsen 2007; The University of South Carolina, Technology Commercialization Office 2020; Labiotech EU 2021; Institute Ruđer Bošković, Hrvatska 2020):

- Researchers from China have developed an intelligent device (microneedle) that can monitor blood glucose levels and, in response, release a biochemical to regulate said levels.
- Turkish scientists have been investigating the relationship between serum lipoprotein(a) levels and red cell distribution width in healthy adult men. Estimates of lipoprotein(a) and red cell distribution width may be helpful to predict the risk for coronary heart disease, heart failure, hypertension, arrhythmias, and stroke in healthy subjects in the future.
- Researchers at Harvard University (United States) have investigated intestinal bacteria and found that such bacteria produce digoxin, which protects the heart.
- Researchers at the University of South Carolina (United States) have innovated a new type of drug based on experimentally validated selective inhibitors of *Trypanosoma cruzi* glucokinase, which offer an alternative to commonly prescribed medications for two diseases of the trypanosome, including American trypanosomiasis (Chagas' disease), which is caused by the *T cruzi* parasite, and African trypanosomiasis (African sleeping sickness), which is caused by the *T brucei* parasite.
- Researchers at the Polytechnic University ofValencia (UPV) and the La Fe Health Investigation Institute (IIS La Fe) (Spain) have developed Moosy 32eNose, a technology (a prototype of an electronic nose) that can help sniff out disease. This new device can help distinguish between patients with Crohn's disease and ulcerative colitis. In addition, the electronic nose could be used to detect pancreatic cancer or to detect microbial contamination of water in the future.

- The Royal Swedish Academy of Sciences has decided to award the 2017 Nobel Prize in Chemistry to three brilliant scientists, namely Jacques Dubochet, Joachim Frank, and Richard Henderson, for developing cryo-electron microscopy for the high-resolution structure determination of biomolecules in solution. This development of cryo-electron microscopy has enabled and improved the imaging of biomolecules, which, in turn, has moved biochemistry into a new era.
- US-based start-up DiPole Materials offers nanofiber manufacturing solutions. It uses its BioPaper technology to provide a gelatin-based three-dimensional fibrous scaffold for three-dimensional cell culture and bioprinting. This solution accelerates the development of novel medical devices while simultaneously improving the process of screening pharmaceutical drugs.
- Researchers from the Ruđer Bošković Institute (Croatia) are working on "CAT Pharma", an innovative platform for developing catalytic reactions to be used in the chemical and pharmaceutical industries to improve said industries' outdated processes. The use of catalytic technologies is directly linked to production savings. The main characteristics include reducing energy consumption and chemical solvents or reagents, increasing production capacity, accelerating production processes, and reducing the amount of chemical waste. Thus, catalytic protocols can make chemistry safer, more economical, and environmentally friendly. The project envisages the development of three categories of catalytic processes that use metal-organic catalysts and biocatalysts.
- BioMed Elements is a Dutch start-up developing orthobiologics or biomaterials for the medical device and cosmetics industry. It uses BioMed Core technology to produce biopolymers into various functional end forms, such as spherical submicron particles. Different materials such as ceramics, tissue, and bone are mixed to produce orthobiological drugs for orthopaedic and dental practices.

Biopharmaceutical innovations make it possible to treat many previously deemed incurable conditions while reducing the economic burden on the health system and increasing the quality of patients' lives. Advances in biopharmaceutical research have had an enormous impact on individuals and the wider community. Improving the health system requires policies that support the R&D system.

In practice, biopharmaceuticals include three main categories, namely genetic engineering drugs, biological vaccines, and biological diagnostic agents, which play a vital role in the diagnosis, prevention, control, and eradication of infectious diseases to protect and improve human health.

Biopharmaceutical technology is recognised as one of the critical strategic technologies in the 21st century. Many countries are designing their development and management policies to support the use of biopharmaceuticals and enable biopharmaceutical companies to acquire competencies in this area.

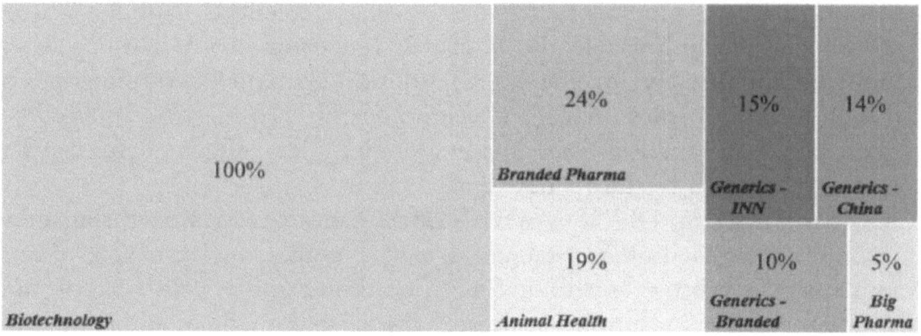

Figure 4.12 Change in the total value of enterprises in each subsector of the biopharmaceutical economy in 2020 (%)

Source: Author

Large centres of the biopharmaceutical industry are concentrated in the United States, Europe, Japan, India, and China (Zhang and Liu 2020).

The innovation surge is probably most visible in the stock market, capitalising on future expected profitability (Figure 4.12) (Torreya 2021).

Since biopharma is a significant sector of the bioeconomy, the following list includes simply some of the innovations in the field (Bio-based Industries Joint Undertaking – BBI-JU 2020; Kulitane 2020; Bio-based Industries Joint Undertaking 2020; Bio-based Industries Consortium 2019):

- The SpiralG project uses seaweed biomass to produce bio-based food dyes, bio-stimulants for plants, functional protein-rich compounds for pet food, and bioactive protein compounds for the pharmaceutical industry.
- Tedre Talu LTD is involved in the production of raspberry seed oil. A local company from south-eastern Estonia produces raspberry seed oil. Raspberry seeds are a by-product of the production of raspberry juice. The company has compared different methods for extracting oil from seeds with bioactive components. After a series of attempts, they were able to find an efficient waste-free method that makes use of 100% of the produced raspberries. They have also developed a new innovative technology for extracting oil from seeds.
- Field and Forest's organic farm regularly initiates and participates in R&D activities focused on plant cultivation methods, biorefineries, and new product development. The company uses unique in-house developed machines and tools to improve the efficiency of production processes. A chemical laboratory equipped with liquid and gas chromatographs, a mass spectrometer, and a spectrophotometer perform a quality assessment. Experimental fields are used to search for the best genetic varieties of plants. Waste is used in biorefineries where new products are also made from by-products. In

addition, traditional technologies are being studied to discover other uses of plants.

- The Arctic Red project (Klosser Innovasjon Ltd) aims to establish a new national fish farming programme for Arctic Char (Salvelinus alpinus) from the family Salmonidae. The programme will accommodate new fish producers with quality roe from a breeding strain that is better adapted to land-based aquaculture and food production. The strain is selected towards faster growth, better use of feed, and later maturation time. The project is establishing the necessary genetic tools for the functioning of a modern fish farming programme, a genetic marker panel for genotyping and species control. For the first time, DNA sequencing is used for fish in Norway.
- The AESTI brand has been researching healthy, natural cosmetic products to solve scalp and hair problems since 2007. The AESTI team joined forces with two Estonian universities to study the use of peat (compost) for deep scalp cleansing. Peat is mainly used for energy and fertilisation. A novel, innovative filtration method was elaborated as part of the research. The lengthy product development process has resulted in the release of natural, organic shampoo. The research process has shown that peat has exceptionally high concentrations of humic substances and minerals. In addition, fulvic acid prolongs the life of skin cells.
- Hovione, a global pharmaceutical company from Portugal, uses new, innovative technologies to produce medicines.
- Bial Biotech is a global innovative pharmaceutical company that produces drugs for neurological diseases. They are experts in lysosomal biology used to produce small innovative molecules to treat neurodegenerative diseases.

Conclusion

In addition to contributing to the industry modernisation, *new emerging sectors* focus on the knowledge- and innovation-based bioeconomy in terms of technology using renewable biological resources for food, materials and energy, and innovation in management and economic models, while enabling commercial success that entails environmental protection, achieving system sustainability, and reducing the impact of climate change.

Start-ups and SME that are often faced with financial risks and long-term investments as well as survival in a well-established fossil fuel market are currently at the forefront of *emerging new sectors*. Developing a sustainable bio-based industry contributes to creating competitiveness, innovation, and new jobs.

Bioenergy, *bio-based chemistry* and *biochemistry*, and *biopharma* have been analysed as *new emerging sectors* in terms of the use of bio-based resources, the development of innovative technologies, and the requirements set by the market.

Significant efforts are being made in the *bioenergy sector* to reduce dependence on conventional energy (i.e., energy from fossil fuels that substantially impact GHGs in the atmosphere) and increase the use of energy from biological resources. In exchange for generating bioenergy, advanced technologies directly or indirectly use organic material or biomass, including plant material and animal waste, as well as waste that is a consequence of human activity (municipal waste). The growth of the bioenergy market is conditioned by the shift to renewable energy sources that are not based on fossil fuels, such as solid biomass, biogas, and so forth. Factors such as increasing energy demand, advances in innovative bioenergy technologies and rising investment in bioenergy, and reductions in the cost of electricity generation from bioenergy plants are expected to boost the market in the analysed periods. However, the growing demand for electric vehicles worldwide is also expected to negatively affect biofuels in the coming years. New, innovative technologies in the bioenergy sector are pushing the boundaries of conventional science. It is estimated that investment in research and innovation in the bioenergy sector could increase added value tenfold. Despite significant advances in innovative technologies that use biofuels to produce bioenergy, the environmental impacts of the production of liquid biofuels for transport, policy costs for their promotion, and possible adverse effects have been severely criticised. Energy from biomass has a growing role in the global energy sector and can significantly contribute to reducing carbon dioxide emissions. However, intensive land-use change can have the opposite effect and significantly increase carbon dioxide emissions.

In the *bio-based chemistry* sector, the development of bio-based chemicals, polymers, and materials depends on new raw materials, innovative technologies, and new economic opportunities. More comprehensive application is limited by several factors such as production costs, security of access to resources, development of the supply sector, the use of renewable sources in production, and recyclable product design. The development of bio-based chemicals is constrained by the lower price of fossil fuels compared to the cost of renewables, trade wars, a lack of policy and legislative coherence, and insufficient or no sanctions for GHG emissions for those who continue to produce fossil fuels. In addition to the primary biofuel production, biorefineries offer opportunities for the development of bio-based products (chemicals and polymers) and materials (fibres, starch derivatives) that are produced as products or as by-products depending on selected technological processes (fermentation, thermochemical conversion, pyrolysis, degassing, acid and ester hydrolysis, chemical conversion, anaerobic digestion, transesterification, etc.). The selection and balance of biorefinery products must be market competitive as this will be crucial for their economic viability.

R&D of new technologies, as well as legal regulations, support the *fermentation products* market growth. There has recently been an increase in the use of *bioplastics* due to the growing awareness of consumers about the harmful effects of conventional products (fossil fuels) on human health and flora and fauna in the ecosystem, consequently making biodegradable plastics that can be re-absorbed into the ecosystem and thus preserve the environment more acceptable. The development of *biopolymers* is intensive, and innovative biopolymers have already been put to use in electronics, construction, paper, packaging, cement, automotive, and other industries. The analyses showed that biopolymers (such as the commercially available PLA biopolymer) have particularly valuable properties and effective use compared to polymers produced from classical chemical processes. In addition to being efficient in application and sustainability, their production has a lower environmental impact because of the use of renewables (biomass). *Lignin* is a by-product of the pulp and paper industry but has wide application in many industries and personal use products. Moreover, the use of lignin represents a significant opportunity to improve lignocellulose biorefineries, while new research and innovations enable higher jet fuel production. *Oleochemicals*, products obtained from natural oils and fats, represent sustainable alternatives to conventional products, are environmentally friendly, and their use contributes to environmental protection. Their growing demand results from public awareness of the need for environmental conservation, acceptable production costs, and rising oil prices. It takes 30 years to implement bio-based chemicals (to reach the full economic potential) if the value of the bio-based product is not lower than the depreciated production costs of the petrochemical counterpart. If the value of bio-based chemical products is lower than the value of depreciated production costs of the petrochemical counterpart, that period is reduced to 10 years. Although new technologies in the bio-based chemistry sector are still emerging, novel bio-based products are helping increase the demand for more sustainable end-use goods. The lack of raw materials (organic chemicals) also contributes to the shift towards new raw materials and production processes. The market will continue to depend on critical economic and political factors and the price of oil. Global market growth forecasts and estimates of CAGRs (%) for bioplastics and biopolymers, fermentation products, oleochemicals, and lignin for the analysed periods show that the highest growth is expected for the bioplastics and biopolymers market, followed by oleochemicals, lignin, and fermentation products' market. However, these forecasts depend on political, economic, and environmental factors.

Technologies used in the *biochemistry sector* that seek to explain the complex chemical reactions in various life forms and understand biological processes are expected to stimulate significant advances in medicine, veterinary medicine, agriculture, and biotechnology (especially protein structure/function and genetic engineering). Innovations in biochemistry are necessary for the application of better solutions that meet new requirements or existing market needs, that is, more efficient solutions for processes, products, services, or innovative ideas in biochemistry but also the possibilities of their application in the production of functional, structural materials and other areas, such as agricultural and waste industries. The analysis of the biochemistry market can be monitored through the biochemistry analysers market, the mass spectrometry market, and the biochemical reagents market, among others. Market analysis assessments in the biochemistry sector have been affected by the Covid-19 pandemic and the economic crisis it has caused. The global biochemistry market is projected to grow significantly (tenfold) over the next few years. These projections include biochemistry applications and processes that positively impact the economy and the environment. In addition, the number of biotech companies in Europe that provide new innovative solutions in the biochemistry sector has been on the rise in recent years. *Biopharmaceuticals* play a crucial role in increasing global health in the biopharmaceutical sector. Research, innovation, and clinical trials in humans and animals are vital to drug development and marketing. Unlike traditional drugs synthesised from chemical processes, most biopharmaceuticals are derived from biological processes, including the extraction from living systems or the production by recombinant DNA technologies or genetic engineering. Emerging sectors include precision medicine and regenerative medicine, which represent entirely new ways of treating disease. Biopharmaceuticals have the potential to play a vital role in increasing global health to achieve the 2030 Sustainable Development Goals. Herbal drug production has enormous potential and could become an essential system for various new biopharmaceutical products. The global biopharmaceuticals market is driven by numerous factors, such as the increase in the elderly population, surge in prevalence of chronic diseases such as cancer and diabetes, and global adoption of biopharmaceuticals. High costs associated with drug development and risky production could prevent the growth of the global biopharmaceuticals market. The pandemic induced by the SARS-CoV-2 virus has immensely increased the demand for biopharmaceuticals and the need for research and innovation in the biopharmaceutical sector. Biopharmaceutical innovations also make it possible to treat many conditions that were previously deemed incurable, which is why the introduction of innovative pharmaceuticals is expected to strengthen the biopharmaceuticals market in the following decade. Improving the health system requires policies that support the R&D system to reduce the economic burden on the health system and increase the quality of patients' lives. Biopharmaceutical technology is recognised as one of the critical strategic technologies in the 21st century because of the advances in biopharmaceutical research that have had an enormous impact on individuals and the wider community.

A critical factor in the realisation of a thriving bio-based economy will be the development of biorefinery systems well integrated into existing infrastructure. Furthermore, to ensure the well-being and survival of humanity, it is necessary to use new innovative bio-based technologies in new emerging sectors, apply redesigned business models aiming to reduce not only environmental and climate risks but also certain new types of risks, such as the Covid-19 pandemic, and design policies to achieve the set goals.

Bibliography

Acme-Hardesty, 2020. Bright future for the bio-based chemical industry. Available from: www.acme-hardesty.com/bio-based-chemical-industry/ [Accessed 27 October 2021].

AD Bioplastics, 2020. AD bioplastics is in the TOP 5 of the best-biobased polymer start-ups in the world. Available from: https://adbioplastics.com/en/2020/09/14/adbioplastics-is-in-the-top-5-of-the-best-biobased-polymer-startups-in-the-world/ [Accessed 19 October 2021].

Adler, L., 2021. Turning industrial waste gases into valuable polyurethanes. Available from: www.carbon4pur.eu/wp-content/uploads/2021/04/2021-03-25_Sustainable-Plastics-Symposium_Carbon4PUR_all-speakers.pdf [Accessed 19 October 2021].

Alhyari, M., 2018. Chemoenzymatic synthesis of polypeptides for use as functional and structural materials. 4th International Conference on Chemical Engineering, September 17–18, 2018, Vancouver, Canada, Abstract paper. Available from: www.longdom.org/proceedings/chemoenzymatic-synthesis-of-polypeptides-for-use-as-functional-and-structural-materials-19721.html [Accessed 1 November 2021].

Allied Market Research, 2018. Bioenergy market expected to reach $217.8 billion by 2030. Available from: www.alliedmarketresearch.com/press-release/bioenergy-market.html [Accessed 27 September 2021].

Allied Market Research, 2021. Mass spectrometry market by technology. Available from: www.alliedmarketresearch.com/mass-spectrometry-market [Accessed 4 November 2021].

Balskus, E. P., and Jacobsen, E. N., 2007. Asymmetric catalysis of the transannular Diels-Alder reaction. *Science*, 317 (5845), 1736–1740. Available from: www.science.org/doi/10.1126/science.1146939 [Accessed 18 November 2021].

Basilisk Self-Healing Concrete 2021. The benefits of self-healing concrete. Available from: www.basiliskconcrete.com/en/ [Accessed 27 October 2021].

Batchelor, S., 2012. New opportunities: Renewable chemicals and biobased products. Available from: http://energyoutlook2012.naseo.org/presentations/Batchelor-Renewable_Chemicals_and-Biobased_Products.pdf [Accessed 21 September 2021].

Billings, A., 2021. Market analysis: Biochemistry and analytical chemistry, London on May 10–13, 2021. *Journal of Biochemistry & Analytical Biochemistry*, 10 (2). Available from: www.longdom.org/open-access/market-analysis-biochemistry-and-analytical-chemistrylondon-on-may1013-2021.pdf [Accessed 4 November 2021].

Bio-based Industries Consortium (BIC), 2019. *Mapping Portugal's bio-based potential*, European Forestry House, Belgium. Available from: https://biconsortium.eu/sites/biconsortium.eu/files/downloads/BIC%20country%20report%20-%20Mapping%20Portugal%E2%80%99s%20bio-based%20potential.pdf [Accessed 26 November 2021].

Bio-based Industries Joint Undertaking – BBI-JU, 2020. 7 years advancing the European bio-based industry. Available from: www.bbi.europa.eu/sites/default/files/media/BBI-JU-impacts-2014-2020_0.pdf [Accessed 26 November 2021].

Bio Explorer, 2018. Top 14 biochemistry news, innovations & breakthroughs in 2017. Available from: www.bioexplorer.net/biochemistry-news-2017.html/ [Accessed 19 October 2021].

Bioenergy Europe, 2021. Joint Statement: Improper Sustainability Framework for Forest Biomass Could Challenge the EU's Renewable Energy Ambitions and Harm the Rural Economy. Available from: https://bioenergyeurope.org/articles/316-joint-statement-improper-sustainability-framework-for-forest-biomass-could-challenge-the-eu-s-renewable-energy-ambitions-and-harm-the-rural-economy.html [Accessed 23 September 2021].

BioProcess Online, 2006. What is biopharmaceutical? Available from: www.bioprocessonline.com/doc/what-is-biopharmaceutical-0001 [Accessed 4 November 2021].

BioSpace, 2021. Antiviral drugs market size to reach USD 77.07 billion in 2028 – Reports and Data. Available from: www.biospace.com/article/antiviral-drugs-market-size-to-reach-usd-77-07-billion-in-2028-reports-and-data/ [Accessed 5 October 2021].

Biotechnology Innovation Organization, 2021. What is biotechnology? Available from: www.bio.org/what-biotechnology [Accessed 21 September 2021].

Bomtempo, J. V., and Alves, F. C., 2014. Innovation dynamics in the biobased industry. *Chemical and Biological Technologies in Agriculture*, 1 (19). Available from: https://

chembioagro.springeropen.com/articles/10.1186/s40538-014-0019-8#citeas [Accessed 11 October 2021].

Bouzidi, A., 2021. Biopharma and a life science ecosystem: Putting the foundations in place for a productive partnership. *European Pharmaceutical Review*. Available from: www.europeanpharmaceuticalreview.com/article/158115/biopharma-and-the-life-sciences-ecosystem-putting-the-foundations-in-place-for-a-productive-partnership/ [Accessed 5 October 2021].

Brightmark, 2021. The Yellowjacket project in renewable natural gas. Available from: www.brightmark.com/work/yellowjacket-project/ [Accessed 1 October 2021].

Bugge, M. M., Hansen, T., and Klitkou, A., 2016. What is the bioeconomy? A review of the literature. *Sustainability*, 8 (7), 691.

Building the Bioeconomy 5th Edition, 2018. National biotechnology industry development strategies globally, Pugatch Consilium, UK. Available from: https://internationalbiotech.org/wp-content/uploads/2018/11/BIO-2018-report-Final.pdf [Accessed 21 September 2021].

Carbios, 2020. Carbios produces first clear plastic bottles from enzymatically recycled textile waste. Available from: www.carbios.com/en/carbios-produces-first-clear-plastic-bottles-from-enzymatically-recycled-textile-waste/ [Accessed 27 October 2021].

Carbon Recycling International, n.d. George Olah renewable methanol plant: First production of fuel from CO_2 at industrial scale. Available from: www.carbonrecycling.is/project-goplant [Accessed 1 October 2021].

Celik, A., and Kilinc, M., 2016. Relation between red cell distribution width and serum lipoprotein (a) in healthy adult men. *Clinical and Medical Biochemistry*, 2 (3), 1–3. Available from: www.longdom.org/open-access/relation-between-red-cell-distribution-width-and-serum-lipoproteina-in-healthy-adult-men-2471–2663–1000120.pdf [Accessed 18 November 2021].

Chen, Y. C., and Yeh, M. K., 2018. *Introductory Chapter: Biopharmaceuticals*. London: IntechOpen. Available from: www.intechopen.com/chapters/62512 [Accessed 4 November 2021].

CISION PR Newswire, 2021a. Global lignin markets report 2021–2031: Future outlook, production, capacities & profiles of 75+ producers. Available from: www.prnewswire.com/news-releases/global-lignin-markets-report-2021-2031-future-outlook-production-capacities–profiles-of-75-producers-301305014.html [Accessed 11 October 2021].

CISION PR Newswire, 2021b. Global oleochemicals market report 2021–2027. Featuring major players from Indonesia, Malaysia, and the Philippines. Available from: www.prnewswire.com/news-releases/global-oleochemicals-market-report-2021-2027-featuring-major-players-from-indonesia-malaysia-and-the-philippines-301369090.html [Accessed 11 October 2021].

Clariant, 2016. Bioethanol made from agricultural residues used for the first time in cleaning agents. Available from: www.clariant.com/en/Corporate/News/2016/02/Bioethanol-made-from-agricultural-residues-used-for-the-first-time-in-cleaning-agents [Accessed 1 October 2021].

Department for Business, Energy and Industrial Strategy, 2019. *Energy Innovation Needs Assessment – Biomass & Bioenergy*. London: Department for Business, Energy and Industrial Strategy. Available from: https://assets.publishing.service.gov.uk/government/uploads/system/uploads/attachment_data/file/845653/energy-innovation-needs-assessment-biomass-bioenergy.pdf [Accessed 15 November 2021].

Dou, J., 2020. Biochemistry 2020 market analysis. *Journal of Molecular and Cellular Biochemistry*, 3 (2). Available from: www.imedpub.com/articles/biochemistry-2020-market-analysis.pdf [Accessed 2 November 2021].

Ductor. Our projects. Available from: www.ductor.com/ [Accessed 1 October 2021].

European Biogas Association, 2018. Seat and Aqualia pioneer biomethane production. Available from: www.europeanbiogas.eu/spain-seat-aqualia-pioneer-biomethane-production/; www.life-methamorphosis.eu/en/el-projecte [Accessed 1 October 2021].

European Commission, 2019a. *Brief on Agricultural Biomass Production*. Brussels: European Commission. Available from: https://op.europa.eu/en/publication-detail/-/publication/40ff478c-4769-11e8-be1d-01aa75ed71a1/language-en [Accessed 23 September 2021].

European Commission, 2019b. Facts and figures on bioenergy in the EU. *EU Science Hub*. Available from: https://ec.europa.eu/jrc/en/science-update/facts-and-figures-bioenergy-eu [Accessed 22 September 2021].

European Commission, 2019c. Renewables annual questionnaire 2017–2021 and historical revisions. Available from: https://ec.europa.eu/eurostat/documents/38154/10015688/Ren-Instructions-2018.pdf/bc81e198-d1fc-4799-bb35-f5f4f252e19b [Accessed 22 September 2021].

EU Science Hub, 2019. *The Future of Bio-based Chemicals in the EU Bioeconomy*. Brussels: European Commission. Available from: https://ec.europa.eu/jrc/en/science-update/future-bio-based-chemicals-eu-bioeconomy [Accessed 4 October 2021].

FAO, 2020. Global forest resources assessment 2020. Available from: www.fao.org/documents/card/en/c/ca9825en/ [Accessed 23 September 2021].

Fortune Business Insights, 2020a. Bioenergy market to hit USD 642.71 billion by 2027; growing need to replace fossil fuels with renewable energy sources worldwide to fuel market expansion. Available from: www.globenewswire.com/news-release/2020/07/20/2064486/0/en/Bioenergy-Market-to-Hit-USD-642-71-Billion-by-2027-Growing-Need-to-Replace-Fossil-Fuels-with-Renewable-Energy-Sources-Worldwide-to-Fuel-Market-Expansion-Fortune-Business-Insights.html [Accessed 27 September 2021].

Fortune Business Insights, 2020b. Market research report, bioethanol market size, share and global trend by feedstock. Available from: www.fortunebusinessinsights.com/industry-reports/bioethanol-market-101076 [Accessed 27 September 2021].

Fortune Business Insights, 2020c. Market research report, biogas market size, share and COVID-19 impact analysis. Available from: www.fortunebusinessinsights.com/industry-reports/biogas-market-100910 [Accessed 27 September 2021].

Fortune Business Insights, 2020d. Market research report, biomethanol market size, share and global trend by source. Available from: www.fortunebusinessinsights.com/industry-reports/bio-methane-market-100272 [Accessed 27 September 2021].

Fortune Business Insights, 2020e. Market research report, energy and power. Available from: www.fortunebusinessinsights.com/industry-reports/biomass-power-generation-market-100746 [Accessed 27 September 2021].

FuelCellsWorks, 2020. UNIST researcher develop a novel biofuel system for hydrogen production from biomass. Available from: https://fuelcellsworks.com/news/unist-researchers-develop-a-novel-biofuel-system-for-hydrogen-production-from-biomasss/ [Accessed 1 October 2021].

Glitsoe, V., Ruckebusch, J. P., and Knap, I., 2015. Innovation in enzyme development, animal feed. Available from: https://en.engormix.com/feed-machinery/articles/innovation-enzyme-development-t36552.htm [Accessed 2 November 2021].

Global Market Insights, 2020. Lignin market size, by product (Kraft Lignin, Lignosulphonates, Low Purity Lignin), by application (Aromatics, Macromolecules), industry analysis report, regional outlook, lignin downstream potential (Vanillin, Carbon Fiber, Phenol, BTX), application potential, price trends, competitive market share & forecast, 2021–2027.

Available from: www.gminsights.com/industry-analysis/lignin-market [Accessed 11 October 2021].

GlobeNewsWire, 2021. Lignin market value to cross $1.14 billion by 2027. Available from: www.globenewswire.com/news-release/2021/09/15/2297611/0/en/Lignin-Market-value-to-cross-1-14-billion-by-2027-Says-Global-Market-Insights-Inc.html [Accessed 11 October 2021].

Golden, J. S., & Handfield, R. B., 2014. *Opportunities in the Emerging Bioeconomy*. Washington, DC: Duke University. Available from: www.biopreferred.gov/files/WhyBiobased.pdf [Accessed 4 October 2021].

Goldstein, D. A., and Thomas, J. A., 2004. Biopharmaceuticals derived from genetically modified plants QJM. *An International Journal of Medicine*, 97 (11), 705–716.

Greenbelt Resources Corporation, 2019. Dairy processor waste solutions: A new era. Available from: www.greenbeltresources.com/dairy-processor-waste-solutions-a-new-era/ [Accessed 1 October 2021].

Green Facts, 2009. Liquid biofuels for transport prospects, risks, and opportunities. Available from: www.greenfacts.org/en/biofuels/l-2/1-definition.htm [Accessed 22 September 2021].

IEA Bioenergy, 2020. Bio-based chemicals – A 2020 update, task 42. Available from: www.ieabioenergy.com/wp-content/uploads/2020/02/Bio-based-chemicals-a-2020-update-final-200213.pdf [Accessed 4 October 2021].

IEA Bioenergy, 2021. IEA tracking clean energy progress – Biofuels/bioenergy. Available from: www.ieabioenergy.com/blog/publications/iea-tracking-clean-energy-progress-biofuels-bioenergy/ [Accessed 15 November 2021].

IFP Energies Nouvelles, 2021. Bio-based chemistry. Available from: www.ifpenergiesnouvelles.com/innovation-and-industry/our-expertise/renewable-energies/bio-based-chemistry [Accessed 1 October 2021].

Industry ARC, 2021. Fermentation products market – Forecast (2021–2026), Report Code: CMR 91891. Available from: www.industryarc.com/Research/Fermentation-Products-Market-Research-501680 [Accessed 11 October 2021].

Institute Ruđer Bošković, Hrvatska, 2020. Review – Razvijamo inovacije za primjenu u kemijskoj, farmaceutskoj i biomedicinskoj industriji. Available from: www.irb.hr/Novosti/Razvijamo-inovacije-za-primjenu-u-kemijskoj-farmaceutskoj-i-biomedicinskoj-industriji [Accessed 22 November 2021].

Journal of Biomedical Sciences, 2019. Market analysis on biotechnology and bioengineering-Summary. *Journal of Biomedical Sciences*, 8, 1–4. Available from: www.jbiomeds.com/biomedical-sciences/market-analysis-on-biotechnology-and-bioengineering.pdf [Accessed 4 November 2021].

Kulitane, I., ed., 2020. The 25 cases for bioeconomy innovations around Baltic Sea Region. Available from: www.jamk.fi/globalassets/tutkimus-ja-kehitys – research-and-development/tki-projektien-lohkot-ja-tiedostot/rdi2club/results/25casesforbioeconomy-eng-26082020-web.pdf [Accessed 26 November 2021].

Labiotech EU, 2021. The 21 European Biotech Companies to watch in 2021. Available from: www.labiotech.eu/best-biotech/european-biotech-companies/ [Accessed 22 November 2021].

Lange, L., Connor, K. O., Arason, S., Bundgård-Jørgensen, U., Canalis, A., Carrez, D., Gallagher, J., Gøtke, N., Huyghe, C., Jarry, B., Llorente, P., Marinova, M., Martins, L. O., Mengal, P., Paiano, P., Panoutsou, C., Rodrigues, L., Stengel, D. B., van der Meer, Y., and Vieira, H., 2021. Developing a sustainable and circular bio-based economy in EU: By partnering across sectors, upscaling and using new knowledge faster, and for the benefit of climate, environment & biodiversity, and people & business. *Frontiers in Bioengineering and Biotechnology*, 8, 619066. Available from: www.frontiersin.org/articles/10.3389/fbioe.2020.619066/full

Lee, S.Y., Sankaran, R., Chew, K.W., Tan, C. H., Krishnamoorty, R., Chu, D.T., and Show, P. L., 2019. Waste to bioenergy: A review on the recent conversion technologies. *BMC Energy*, 1 (4). Available from: https://doi.org/10.1186/s42500-019-0004-7

Market Data Forecast, 2021. Global biopharmaceuticals market size, share, trends, growth, Covid-19 impact analysis report – Segmented by product type, therapeutic type, and region – Industry forecast (2021–2026). Available from: www.marketdataforecast.com/market-reports/bio-pharmaceuticals-market [Accessed 15 November 2021].

Market Research Engine, 2021. Lignin market research report, global lignin market research report – Industry analysis, size, share, growth, trends and forecast 2021–2026. Available from: www.marketresearchengine.com/reportdetails/global-lignin-lignosulfonates-kraft-lignin-market-report [Accessed 11 October 2021].

Markets and Markets, 2015. Oleochemicals market by type (Fatty Acid, Fatty Alcohol, Glycerin, and Others), by application (Pharmaceutical & personal care, Food & beverages, Soaps & detergents, Polymers, and Others), by region (North America, Europe, Asia-Pacific, and Rest of the World) – Trends and forecasts to 2019. Report code: CH 3121. Available from: www.marketsandmarkets.com/Market-Reports/oleochemicals-market-235516809.html?gclid=CjwKCAiA4veMBhAMEiwAU4XRr6LD9-0O34lw1HKz1nCp0bOuKbe7BQKpdmhYayHp6wETTChIMV_59hoCH_oQAvD_BwE [Accessed 11 October 2021].

Markets and Markets, 2021a. Bioplastics & biopolymers market by type (non-biodegradable/bio-based, biodegradable), end-use industry (packaging, consumer goods, automotive & transportation, textiles, agriculture & horticulture), region – Global forecast to 2026. Available from: www.marketsandmarkets.com/Market-Reports/biopolymers-bioplastics-market-88795240.html?gclid=CjwKCAiA4veMBhAMEiwAU4XRr9F422BtHp19FdSyJBXyvDw3arv1QmALDHFh7k0bzNVPREROPpAJBRoCGCAQAvD_BwE [Accessed 11 October 2021].

Markets and Markets, 2021b. Oligonucleotide synthesis market by product. Report code BT 2680. Available from: www.marketsandmarkets.com/Market-Reports/oligonucleotide-synthesis-market-200829350.html [Accessed 15 November 2021].

Miranda, C., Ruiz-Cabello, E., and Camara Hurtado, M., 2020. Food biopharmaceuticals as part of sustainable economy: Edible vaccines case study. *New Biotechnology*, 59, 74–79. Available from: www.sciencedirect.com/science/article/pii/S1871678420301503 [Accessed 4 November 2021].

Mordor Intelligence, 2020a. Biochemical reagents market – growth, trends, COVID-19 impact, and forecasts (2021–2026). Available from: www.mordorintelligence.com/industry-reports/biochemical-reagents-market [Accessed 15 November 2021].

Mordor Intelligence, 2020b. Bioenergy market – Growth, trends, Covid-19 impact, and forecasts (2021–2026). Available from: www.mordorintelligence.com/industry-reports/bioenergy-market [Accessed 15 November 2021].

NEWFERT, 2018. Nutrient recovery from bio-based waste for fertiliser production. Available from: www.bbi.europa.eu/projects/newfert [Accessed 27 October 2021].

OECD/FAO, 2019. *OECD-FAO Agricultural Outlook 2019–2028*. Paris: OECD Publishing/Rome: Food and Agriculture Organization of the United Nations. Available from: https://doi.org/10.1787/agr_outlook-2019-en [Accessed 22 September 2021].

Office of Energy Efficiency & Renewable Energy, 2016. *3 Breakthroughs Show How Bioenergy Innovations Are Energizing the Energy Landscape*. Washington, DC: US Department of Energy. Available from: www.energy.gov/eere/articles/3-breakthroughs-show-how-bioenergy-innovations-are-energizing-energy-landscape [Accessed 15 November 2021].

Patel, M., Crank, M., Dornburg, V., Hermann, B., Roes, L., Hüsing, B., Overbeek, L., Terragni, F., and Recchia, E., 2016. *Medium and Long-term Opportunities and Risks of the Biotechnological Production of Bulk Chemicals from Renewable Resources – The Potential of White Biotechnology*. The BREW

Project, under European Commission. Utrecht: Utrecht University. Available from: www.researchgate.net/publication/27704442_Medium_and_Long-Term_Opportunities_and_Risks_of_the_Biotechnological_Production_of_Bulk_Chemicals_from_Renewable_Resources-The_BREW_Project [Accessed 4 October 2021].

Philip, J. C., Ritchie, R. J., and Allan, J. E. M., 2013. Biobased chemicals: The convergence of green chemistry with industrial biotechnology. *Trends in Biotechnology*, 31 (4), 219–222.

Plug and Play, 2021. Profitable carbon reduction: Leveraging emerging technologies. Available from: www.plugandplaytechcenter.com/resources/profitable-carbon-reduction-leveraging-emerging-technologies/ [Accessed 21 September 2021].

PRNewswire, 2021. Global biochemistry analyzers market to reach $4.7 billion by 2026. Available from: www.prnewswire.com/news-releases/global-biochemistry-analyzers-market-to-reach-4-7-billion-by-2026–301367529.html [Accessed 4 November 2021].

Reid, W. V., Ali, M. K., and Field, C. B., 2020. The future of bioenergy. *Global Change Biology,* 26, 274–286. Available from: https://onlinelibrary.wiley.com/doi/epdf/10.1111/gcb.14883

Renewable Energy-Recast to 2030 (RED II), 2019. EU science hub, the European Commission's science and knowledge service. Available from: https://ec.europa.eu/jrc/en/jec/renewable-energy-recast-2030-red-ii [Accessed 22 September 2021].

Research and Markets.com, 2021a. Bioenergy market – Growth, trends, and forecasts (2020–2025). Available from: https://finance.yahoo.com/news/global-bioenergy-market-report-2020-164000238.html?guccounter=1&guce_referrer=aHR0cHM6Ly93d3cuZ29vZ2xlLmNvbS8&guce_referrer_sig=AQAAAL0QVQQdoDYQQ_0K4aXGe5RAfYgAAnhJ8RPRIC9ek_ftljYystXuSMF52eh0olQ4L1QSlSzFfQqrdLx_jkYCcxWUnwG0YrraHzepKfbEXzVtHC6_44qf1T7C_4mQc8CU5C0skpeMy29WpsZ_KzUZbyAfJ7I1vdJbMqg-0ZQxKVXt [Accessed 27 September 2021].

Research and Markets.com, 2021b. Global bioenergy market report 2020–2025: Bioenergy installed capacity GW & bioenergy power generation TWh. Available from: https://finance.yahoo.com/news/global-bioenergy-market-report-2020-164000238.html?guccounter=1&guce_referrer=aHR0cHM6Ly93d3cuZ29vZ2xlLmNvbS8&guce_referrer_sig=AQAAAL0QVQQdoDYQQ_0K4aXGe5RAfYgAAnhJ8RPRIC9ek_ftljYystXuSMF52eh0olQ4L1QSlSzFfQqrdLx_jkYCcxWUnwG0YrraHzepKfbEXzVtHC6_44qf1T7C_4mQc8CU5C0skpeMy29WpsZ_KzUZbyAfJ7I1vdJbMqg-0ZQxKVXt [Accessed 27 September 2021].

Research and Markets.com, 2021c. Global bioenergy market trends and forecasts 2021–2025: Growing need for renewable clean fuel drives growth. Available from: www.prnewswire.com/news-releases/global-bioenergy-market-trends-and-forecasts-2021-2025-growing-need-for-renewable-clean-fuel-drives-growth-301361563.html [Accessed 27 September 2021].

Schepartz, A., 2018. Introducing the "Future of Biochemistry" special issue. *Biochemistry*, 57 (1), 1–8. Available from: https://pubs.acs.org/doi/pdf/10.1021/acs.biochem.7b01259 [Accessed 1 November 2021].

SelectUsa, 2021. The biopharmaceutical industry in the United States: Overview. Available from: www.selectusa.gov/pharmaceutical-and-biotech-industries-united-states [Accessed 4 November 2021].

Smith, J., 2019. *First Industrial Plant to Turn Waste into Chemicals Will Open in France*. Berlin: Labiotech. Available from: www.labiotech.eu/more-news/afyren-recycling-organic-waste-chemical/ [Accessed 19 October 2021].

Spekreijse, J., Lammens, T., Parisi, C., Ronzon, T., & Vis, M., 2019. JRC science for policy. Insights into the European market for bio-based chemicals, JRC, European Commission.

Available from: https://publications.jrc.ec.europa.eu/repository/handle/JRC112989 [Accessed 4 October 2021].

The State of the World's Forests, 2020. Food and Agriculture Organization of the United Nations Rome. Available from: www.fao.org/3/ca8642en/ca8642en.pdf [Accessed 23 September 2021].

Torreya, 2021. Biopharmaceutical sector update. Available from: https://torreya.com/publications/biopharma-market-update-torreya-2021-01-11.pdf [Accessed 26 November 2021].

Trzmielak, D. M., and Kochańska, E., 2021. Innovation in bioenergy. In E. Koukios and A. Sacio-Szymańska, eds. *Factors Affecting Innovation in Biofuels in Bio#Futures*. Berlin: Springer Nature, 413–433. Available from: https://link.springer.com/chapter/10.1007/978-3-030-64969-2_19 [Accessed 15 November 2021].

Univar Solutions, n.d. Oleochemicals used in many industries, Applications. Available from: www.chemicalassociates.com/industries-served.html [Accessed 4 October 2021].

The University of South Carolina, Technology Commercialization Office, 2020. Innovations in chemistry and biochemistry, Ref # 01133 therapeutic inhibitors of hexokinase and glucokinase. Available from: https://sc.edu/about/offices_and_divisions/technology_commercialization/for_industry/browse/chemistry_and_biochemistry/index.php [Accessed 22 November 2021].

Valmet, 2021. LignoBoost – Lignin from pulp mill black liquor. Available from: www.valmet.com/pulp/other-value-adding-processes/lignin-extraction/ [Accessed 19 October 2021].

Vehvilainen, A., Dupont Inglis, J., Kulišić, B., and Barrett, P., 2021. *Deploying the Bioeconomy in the EU: A Framework Approach for Bioeconomy Strategy Development*. Brussels: European Commission. Available from: https://bioeast.eu/wp-content/uploads/2021/07/PSF-Final-Report_Deploying-the-Bioeconomy-in-the-EU_A-Framework-approach-for-bioeconomy-strategy-development_July-2021.pdf [Accessed 21 September 2021].

Wei, C., Rui, T., Can, X., Bryant, C.Y., Guohao, W., Yijing, L., Qianqian, N., Fuwu, Z., Zijian, Z., Jingjing, W., Gang, N., Ying, M., Liwu, F., and Xiaoyuan, C., 2017. Microneedle-array patches loaded with dual mineralized protein/peptide particles for type 2 diabetes therapy. *Nature Communications*, 8 (1777). Available from: www.nature.com/articles/s41467-017-01764-1 [Accessed 18 November 2021].

Wood Energy, 2021. Food and Agriculture Organization of the United Nations. Available from: www.fao.org/forestry/energy/en/ [Accessed 27 September 2021].

ZELCOR Project, 2021. Zero waste ligno-cellulosic biorefineries by integrated lignin valorisation. Available from: https://zelcor.eu/ [Accessed 27 October 2021].

Zhang, K., and Liu, W., 2020. The current status, trend and development strategies of Chinese biopharmaceutical industry with a challenging perspective. *Sage Open Journals* January–March: 1–16. Available from: https://journals.sagepub.com/doi/full/10.1177/2158244020901529 [Accessed 26 November 2021].

5 Holistic approach to bioeconomy monitoring and evaluation

Sanja Tišma

Introduction

To ensure long-term, successful bioeconomy development, it is essential to accurately measure, monitor, analyse, and evaluate the outcomes of strategies and plans that the international community, individual countries, or regions adopt in accordance with their specific goals and landscape. When monitoring the bioeconomy development, it is crucial to include all four pillars of sustainability, namely the environmental, economic, social, and institutional/managerial. In addition, the economic dimension should certainly include technological advances. In this way, the evaluation of the bioeconomy is given a holistic aspect.

Holistically monitoring and evaluating the bioeconomy aims to provide comprehensive information on events in the field of bioeconomy and the impact of these events on the landscape in which development processes take place. A comprehensive insight into changes in the field of bioeconomy represents a framework for improved stakeholder outreach and ultimately a transparent and clear process of making development decisions and adopting related public policies. It is important to emphasise that any holistic assessment of the bioeconomy development must consider the indicators of the specific geographical/territorial area (the place perspective), the time perspective, and the diversity of stakeholders (the people perspective) and their often different insights and sometimes conflicting interests in terms of developments in the field of bioeconomy.

The potential and ultimately desirable framework for monitoring the progress of the bioeconomy is related to the 17 Sustainable Development Goals (SDGs) of the 2030 Agenda for Sustainable Development (UN 2015). The goals emphasise a holistic approach to achieving sustainable development for all. They are related to achieving no poverty, zero hunger, good health and well-being, quality education, gender equality, clean water and sanitation, affordable and clean energy, decent work and economic conditions, industry, innovation and infrastructure, reduced inequality, sustainable cities and communities, responsible consumption and production, climate action, life below water, life on land, peace and justice, and strong institutions and partnerships.

The 231 indicators monitoring the progress on achieving SDGs could partially be used as indicators of the bioeconomy development. The main

DOI: 10.4324/9781003223733-6

advantage of using this set of indicators for achieving the SDGs is that they are already collected and processed at the global level and are internationally comparable (United Nations 2015). However, it is necessary to rethink bioeconomic indicators given that the bioeconomy as a development framework has not been considered sustainably and holistically (i.e., through the place, time, and people perspective, while including economic, social, environmental, and institutional impacts on society). For example, while analysing the environmental dimension of bioeconomy development, it is possible to take notice of several benefits for sustainable economic and social development in traditional sectors of the bioeconomy, namely agriculture, forestry, and aquaculture. In addition, it is possible to observe potential risks that are significantly influenced by the people perspective that might adversely affect environmental factors in the future. There are several instances in which bioeconomy development should be monitored or researched in a multidisciplinary manner. For example, these include the increased competition between the food and fuel production, land-use change, use of marginal lands with adverse effects on biodiversity, GHG emissions, and biomass production that puts pressure on surrounding natural ecosystems, especially in terms of water use and soil quality and degradation (Pfau *et al.* 2014). In addition, until 2000, the sustainability in the bioeconomy was mainly based on observing purely economic effects such as GDP growth, and so on, which is now considered an entirely inadequate system of metrics for assessing general, global well-being over time (Stiglitz *et al.* 2009). Some of the key indicators specific to monitoring bioeconomy development should also focus on monitoring and evaluating new, growing sectors of the bioindustry, biomedicine, biopharmacy, bioenergy, bioinformatics, and the like, indicating the need to describe and monitor new cross-sectoral links and create new value chains.

Recent scientific research and expert analyses have resulted in a wide range of parameters and indicators related to the bioeconomy. Researchers and experts are faced with the challenge of overcoming the complexity and vast amount of information with the help of metadatabases that collect and process data for different purposes, including adapting and aggregating data for holistically monitoring and evaluating bioeconomy development. It is, therefore, necessary to think of new models or adapt existing methods for data collecting and processing and comprehensively draw up the framework for decision-makers. Multisectoral indicator matrices, life-cycle analyses, and input-output tables are some methods that enable decision-making based on a broad and multidisciplinary range of evidence resulting from summing a large number of visible variables (indicators) into a holistic framework as a basis for development decision-making in the field of bioeconomy.

Indicators for monitoring the bioeconomy development

The term "indicator" denotes the representative values of an observed case. Indicators are measures based on data that show the state or change in the state of a system. Each indicator has its specific purpose and is often taken from reporting systems used to measure development policies and used in international reporting. Indicators may also be set out ad hoc to monitor a specific aspect of development that is in the current decision-makers' focus of interests. There are core indicators, which describe crucial, globally recognised trends, process indicators, which are more sophisticated in terms of a certain level of compliance and are sector-specific, and generic or systemic indicators that require a higher level of value assessment given the higher level of complexity of issues they address (Beg 2019).

Quantitative indicators are measurable indicators obtained by documentary analysis and desk research from available sources. These are data that can be collected directly from companies involved in the project or from official data (i.e., existing statistical and other databases, strategic documents, and projects). It is advisable to look at the data for specific quantitative indicators for one or more years before the intervention to draw correct conclusions about the intervention's positive (or negative) impacts concerning the observed indicator. Thus, ex-post evaluation allows the comparison of ex-ante forecasts and observed impacts to assess bioeconomy effectiveness and long-term contribution.

The people perspective aims at gathering opinions and attitudes through stakeholder surveys and other tools (such as interviews, focus groups, public hearings, public insights). Trying to get a hold of them is an opportunity to involve relevant stakeholders. Besides, it is possible to collect additional information that sheds light on participants' satisfaction, the relevant mechanisms responsible for the impact of the intervention, and general feedback to adjust and improve the intervention through surveys, interviews, focus groups, and/or case studies.

In addition, when choosing indicators, it is necessary to respect the criteria of relevance, feasibility (availability), credibility, clarity, and comparability. Relevance means that a particular indicator provides adequate information and response for a specific outcome. Feasibility (availability) refers to the ways and possibilities of collecting specific data or information. The credibility of the information and its reliability depend on the credibility of the data provider used to create the indicator. Clarity is a criterion that indicates the comprehensibility of data and information concerning the user's indicator-related knowledge and skills. Finally, comparability is a criterion that shows changes over time and results that can be compared based on the geographical area they cover.

Using indicators to monitor and evaluate the bioeconomy is still at an early stage and faces several methodological challenges (Karvonen *et al.* 2017). Bioeconomic indicators can be defined as measures of the presence and magnitude of a particular current phenomenon, signs of a future situation or issue, a measure of risk, or the possible need for action and means of identifying and measuring performance in the bioeconomy. The specificity of these indicators depends on the scope and complexity of the bioeconomic area.

The most widely used indicators are economic, social, and environmental indicators, and recently also institutional or management indicators. Economic indicators for the bioeconomy development mainly refer to the generated total turnover, share in gross domestic product, and job creation. These are highly generic indicators that rarely describe the processes in the field of bioeconomy. However, the evaluation and monitoring of the bioeconomy are expected to shift from less generic towards more informative indicators, such as improving value chains compared to traditional value chains, or, for example, management styles and methods encouraging the bioeconomy development in local conditions. Contributing to creating new jobs can also be perceived as a social indicator. In addition to this indicator, several other indicators have been developed that describe the impact of the bioeconomy development on society. Unlike economic indicators that are primarily quantitative and can be collected and generated from publicly available statistical databases, social indicators are primarily qualitative and depend on the stakeholders' attitudes (Diaz-Chavez 2014). For example, value-added and employment rates are economic indicators of the impact of the bioeconomy on the local or national economy. In contrast, social indicators relate to analysing working conditions or the quality of employment benefits (Diaz-Chavez 2014, pp. 11–12). The Human Development Index represents an exciting and increasingly used social indicator of a high level of aggregation. It comprises three social dimensions: a decent living standard, health, and knowledge. It is also possible to use the Gini index to evaluate income inequality, typically used to monitor the SDGs (Diaz-Chavez 2014). Social indicators are used to analyse a particular social phenomenon in society as a whole (Acevedo *et al.* 2015). They have become popular with the emergence of concerns about biofuel production in developing countries. However, they are more challenging to monitor and quantify, as they entail in-depth studies, such as household surveys, which are lengthy and expensive (Wesseler and von Braun 2017). In the debate on the bioeconomy, social and economic indicators are often given greater attention than environmental indicators, which are more challenging to monitor and more demanding to process and interpret. These indicators are often directly related to monitoring the agriculture and forestry sectors and less to aquaculture. However, interesting ecological indicators have been established over the last few years to monitor bioeconomy development. These include crop area and yield of biotech crops (canola, soy, corn, sugar sorghum, sugar beet, etc.), number of plant varieties created using biotechnology methods, increasing use of biological control in plant growing, share of agricultural waste recycled with biotechnology methods, area of organic farming, area of fast-growing forest plantations, increasing use of biological drugs in forestry, or total biomass production. Given the relatively new concept of this sector, the institutional framework that ties to the bioeconomy and managing knowledge and skills are crucially important for this sector. Therefore, the indicators to be developed in this area should contribute to creating a broader picture and be strongly supported by decision-makers.

Regardless of the type of indicators for the development of the bioeconomy, it is essential that they be structured as follows:

- Each indicator is linked to relevant bioeconomy development goals at the global, national, or regional/local level.
- Each indicator is mapped to an addressed sustainability pillar: economic, social, environmental, and technological, and institutional and managerial.
- Where possible, each indicator is defined according to the relevant source of biomass to which it refers: agriculture, forestry, fisheries, and aquaculture, new bioeconomic sectors such as bioenergy, biomedicine or biopharmacy, nanotechnology, bioinformatics, and so on.
- Where relevant, each indicator will be mapped according to the appropriate step in the value chain, from natural capital stocks, collection (harvesting), production to processing, use, recycling, and disposal.

Indicators of bioeconomy development are important for monitoring the development of bioproducts, the extent of use of biotechnology, and its impact on human health and well-being. Recent global achievements and scientific discoveries in natural and social sciences challenge scientists to link the future evolution of humanity to advances in biotechnology that, together with new technologies such as nanotechnology and bioinformatics, can create a solid foundation for positive economic, political, and social changes. Existing global problems, such as depleted natural resources, supplying the ever-growing population with food and pharmaceuticals, and environmental pollution, have led to searching for alternative chemical raw materials and technologies and replacing the traditional production with bio-based products and bioprocesses. Thus, monitoring and evaluating the bioeconomy, which includes all of the areas mentioned earlier, is even more significant for the evolution of humankind. It is essential to point out that bioeconomy indicators can assess various aspects of development, such as success, productivity, and efficiency, as well as declining trends and uncertainty in the bioeconomy (Egenolf and Bringezu 2019). According to Idrus and Al-Amin (2016), indicators can help assess, quantify, and classify key drivers of progress and innovation in the bioeconomy. They can assess success using measurable benchmarks and serve as criteria in developing strategic reference points. Indicators are mainly used for policy-oriented outcomes and inputs, often providing an empirical reference point to assess the drivers of the bioeconomy and indicate its position on the right or wrong path.

An indicator of the current situation or an assessment of the initial value represents good starting points for analysing the sustainable use of bioresources in terms of the efficiency of the bioeconomy.

The following sustainable development goals are fundamental and directly related to the development of the bioeconomy: zero hunger (SDG 2) through promoting sustainable agriculture; affordable and clean energy (SDG 7) through the use of green energy; industry, innovation, and infrastructure (SDG 9) by fostering innovation, especially the development of new technologies in medicine, pharmacy, energy, and the like, in which the bioeconomy is increasingly important; climate action (SDG 13); life below water (SDG 14) by conserving and sustainably using the oceans, seas, and marine resources; life on land (SDG 15) by protecting and sustainably using terrestrial ecosystems, forests, and halting land degradation and biodiversity loss; and partnerships for the goals (SDG 17) related to strengthening the means of implementation and revitalising the global partnership for sustainable development (Calicioglu and Bogdanski 2021).

In addition to the developed and generally accepted indicators for monitoring the achievement of 17 SDGs, monitoring and evaluating bioeconomy development further requires the development and elaboration of quantitative indicators through a comprehensive value chain that includes bio-based resources, environmental goods and services, industrial transformation processes, circularity, and sustainability. In other words, it is necessary to carefully evaluate sustainable resources of raw materials, their scope, and different existing and potential future uses as a basis for monitoring the bioeconomy. The Food and Agriculture Organization of the United Nations (FAO 2016), the European Commission, and its Joint Research Centre (JRC) have made significant strides in establishing indicators for monitoring the bioeconomy. In 2019 (Bracco et al. 2019), they initiated joint research based on a holistic view of indicators for monitoring the bioeconomy. As a result, the FAO presented indicators that gravitate towards a holistic approach, listing them through three perspectives: the environment, economy, and society. At the same time, JRC developed highly generic indicators and provided policymakers with crucial information clearly and concisely. However, they did not include the technological aspect and breakthroughs and have limited the steps towards new sectors such as biomedicine, biopharma, and bioenergy that are increasingly linked to the bioeconomy and are crucial for further growth and development globally.

In general, defining and selecting indicators relevant to monitoring the long-term effects of the bioeconomy on society should entail the following principles (Diaz-Chavez *et al.* 2016):

- Assessing which indicators are relevant for holistic monitoring of the bioeconomy to reduce the administrative burden
- Automating processes as much as possible by using information technology tools to shorten data collection and processing times
- Using common reporting standards to increase interoperability and facilitate the exchange of data in the context of a multisectoral view of bioeconomy
- Making maximum use of indicators collected in other available systems such as EUROSTAT, etc.
- Being transparent to stakeholders, making monitoring findings publicly available, and providing "open data" to all interested in monitoring the bioeconomy

For example, traditional economic indicators such as transport and jobs are already providing an important insight into the European Union (EU) bioeconomy as they highlight that the bioeconomy represents about 9% of the overall EU economy (Ronzon *et al.* 2020).

The thematic report Sustainable Agriculture, Forestry and Fisheries in the Bioeconomy – A Challenge for Europe (European Commission 2015) explores the interactions between primary sectors and the bioeconomy. Describing the state of the bioeconomy in detail, the report discusses key trends and policies affecting the bioeconomy sectors. It also presents scenarios on what could happen with the development of the bioeconomy paradigm within the fundamental constraint of sustainability. These scenarios, BIO-MODESTY, BIO-BOOM, and BIO-SCARCITY, depend on the alternative futures defined by the two main uncertainties, namely the demand growth for biomass, materials and energy, and supply growth in biomass. The FAO is coordinating global efforts to develop international guidelines for the sustainability of the bioeconomy under the conditions of simultaneously achieving the bioeconomy goals implemented according to the principles, namely food first, sustainable yields, cascading approach, circularity, and diversity. Countries could use these to measure the sustainability of their bioeconomy strategies and monitor the achievement of economic, social, and environmental goals and priorities (Bracco *et al.* 2019).

Indicators that would monitor the development of the bioeconomy according to the methodology proposed by Bracco et al. (2019) included two key approaches in the selection of indicators and product level. Both approaches presupposed strong stakeholder engagement, attention to sustainability issues, the relationship between indicators and pre-established strategic goals, clearly defined data sources, methods for data collection, and assessment methodology. In addition, both approaches imply transparency and effective communication of the results. The key differences refer to the content of the indicators. The territorial approach (at a local, regional, national, or international level) includes several indicators pointing to sectoral or development public policies, while product level indicators relate to specific economic activities (primarily agriculture, forestry, and aquaculture) and monitoring the development of the bioeconomy in those areas. The disadvantage of this approach in relation to holistically monitoring and evaluating the bioeconomy refers to the dichotomy of viewing the bioeconomy as a green development paradigm and as a production process whose central part is a green product rather than overall development.

The 2019 FAO approach also distinguishes between direct, indirect, and proxy indicators, which differ depending on the extent of the indicator's connection to the geographical area or product monitored as an indicator in the development of the bioeconomy. Direct and indirect indicators are generally easily measurable, but proxy indicators are more generic and indicate development trends. This approach is used in monitoring and evaluating the EU Common Agricultural Policy, which is sector based and also deviates from the holistic approach towards the bioeconomy evaluation and prevents the transfer of good practice to other sectors that also contribute to development of the bioeconomy.

A critical step relating to the holistic monitoring and evaluation of the bioeconomy in the FAO approach refers to the creation of the so-called footprints of indicators. These are composite indicators that integrate both the territorial and product base approach, thus enabling a broad assessment of the multidimensional nature of the bioeconomy while also monitoring and evaluating its local dimension. Carbon footprint is an example of such an indicator widely used today. Here, one of the key steps for society as a whole would be, for example, green GDP.

Indicators at the territorial level (value-added, exports, investments, employment, productivity) are exclusively related to strategic development goals. For example, suppose a strategic development goal refers to mitigating and adapting to climate change. In that case, indicators could include public financial support to climate mitigation measures or investment in research and innovation. If, for example, a strategic goal at the national, regional, or local level refers to increasing employability, indicators could include the number of employees in rural and urban areas. However, this approach lacks higher-level generic indicators such as the Global Bioenergy Partnership, which proposes 24 generic indicators related to the development of the bioeconomy that should be measurable at the global level.

Indicators respond to individual impacts and development impacts such as food security, domestic biomass production, land-use rights, or the rights to dispose of natural resources. In addition, they are easily collated to the SDG bioeconomy targets and indicators.

In contrast, indicators at the product level are mainly related to production standards, certificates, and labels in the agriculture, forestry, and aquaculture sectors, through the biomass value chain. They range from biomass production and collection, biomass and bioproducts processing and circularity, to the end of the life cycle. While this provides for strong stakeholder involvement in selecting indicators, the monitoring provider decides which indicators will be taken into account as key performance indicators for evaluating the bioeconomy. This approach also lacks a clear time perspective, thereby precluding the holistic approach to assessing the bioeconomy. Furthermore, the choice of indicators is largely dependent on the target sector (e.g., marine environment or forestry and the like), which does not provide insight into a multisectoral perspective.

The FAO methodology has identified a range of indicators of varying levels of complexity useful for the product value chain approach. They are also linked to the strategic goals of specific key sectors, emphasising traditional sectors such as agriculture, forestry, and aquaculture. Other sectors relevant to the bioeconomy, such as paper, biomaterials, biopharmaceuticals, and bioenergy, have also been identified. However, a small number of indicators have been identified, especially generic ones that could be useful for the holistic branding of the bioeconomy. Some of the more interesting generic indicators that should be included in the holistic evaluation are blue water footprint, food stability, ozone depletion potential, acidification, climate regulation potential, and the like.

The 2019 FAO Report provides an overview of a range of interesting base product-based indicators for the primary sector that can undoubtedly be useful for monitoring further bioeconomy development. The problem lies in the availability and thus the comparability of data at the product level. Data availability and gathering process at the level of bio-based products, and hence bio-based value chains, is still in its infancy. The data are often found in scientific research but are not available in official, publicly available statistical databases. Thus, their usefulness for the holistic evaluation of bioeconomy development is largely limited. A holistic approach to monitoring and evaluating the bioeconomy would in some ways involve connecting territorial and product value-chain approaches with an added dimension of time and people perspective through a multisectoral perspective of sustainability (environmental, economic – including technological, social, and institutional).

A significant step forward in designing a global system for holistic monitoring of bioeconomy evaluation is contained in the recent document, "Implementation of the EU Bioeconomy Monitoring System Dashboards" (Kilsedar *et al.* 2021), which summarises the views of both FAO and EC-JRC on the bioeconomy monitoring system. This proposal is based on the key strategic goals for the bioeconomy, namely food security, sustainability of natural resources, non-renewable resources, climate change, and new jobs, and is divided into individual priorities and indicators that are added and further developed each year.

Table 5.1 Comparative overview of possible goals, criteria, and indicators for monitoring the bioeconomy – SDGs, FAO, EC/JRC

SDG (2015) (United Nations, 2015)	*FAO (2019) (Bracco et al. 2019)*	*EC/JRC (2022) (Melim-McLeod et al. 2022)*
SDG 2: Zero Hunger – Volume of production per labour unit by classes of farming/pastoral/forestry enterprise size – Proportion of agricultural area under productive and sustainable agriculture – Number of plant genetic resources for food and agriculture – Government expenditure index – Agricultural export subsidies	Economic criteria: – Economic development – Inclusive economic growth – Resilience of the rural and urban economy – Risk, monitoring, and accountability systems – Knowledge generation and innovation – Local economies – Consumption/supply – Market mechanisms and policy coherence	Food security: – Availability – Access – Utilisation – Stability – Economic impact of trade – Environmental impacts of trade – Social impacts of trade

(*Continued*)

Table 5.1 (Continued)

SDG (2015) (United Nations, 2015)	*FAO (2019) (Bracco et al. 2019)*	*EC/JRC (2022) (Melim-McLeod et al. 2022)*
SDG 7: Clean Energy – Proportion of population with primary reliance on clean fuels and technology – Renewable energy share in the total final energy consumption (bioenergy share)	Environmental criteria: – Sustainable intensification – Biodiversity – Climate change – Water quality and quantity – Land, soil, forests, and marine environments – Resilience of producers, communities, and ecosystems – Resources efficiency, waste prevention, and waste reuse – Food loss and waste	Natural resources sustainability: – Environmental quality – Ecosystem attributes – Species diversity and abundance – Conservation status – Pressures from forestry – Pressures from marine fisheries – Pressures from freshwater fisheries – Pressures from agriculture – Provisioning services – Regulating services – Cultural services
SDG 9: Industry, Innovation, and Infrastructure – Proportion of the rural population who live within 2 km of an all-season road – Research and development expenditure as a proportion of GDP (bioeconomy related) – Proportion of medium- and high-tech industry value added in total value added (bioeconomy related)	Social criteria: – Food security and nutrition – Rights to natural resources – Food safety, disease prevention, human health – Sustainability of urban centres – Policies, regulations, and institutional setup – Inclusion and engagement/information – Existing knowledge/sound technologies – Cooperation, collaboration, and sharing	Non-renewable resources: – Resource efficiency – Energy efficiency – Biogenic waste prevention – Food waste minimisation – Consumption footprint – Consumption and demand – Production – Reduced dependence on non-renewables – Economic impacts of trade – Environmental impacts – Social impacts of trade – Well-being of urban dwellers
SDG 14: Life Below Water – Proportion of economic zones managed using ecosystem-based approaches – Proportion of fish stocks within biologically sustainable levels – Sustainable fisheries as a proportion of GDP – Number of countries making progress in ratifying . . . the United Nations Convention on the Law of the Sea, for the conservation and sustainable use of the oceans and their resources		Climate change: – Mitigation – Adaptation – Resilience in urban areas

(*Continued*)

Table 5.1 (Continued)

SDG (2015) (United Nations, 2015)	*FAO (2019) (Bracco et al. 2019)*	*EC/JRC (2022) (Melim-McLeod et al. 2022)*
- SDG 15: Life on Land – Forest area as a proportion of total land area – Proportion of important sites for terrestrial and freshwater biodiversity that are covered by protected areas, by ecosystem type – Progress towards sustainable forest management – Coverage by protected areas of important sites for mountain biodiversity – Mountain Green Cover Index – Official development assistance and public expenditure on conservation and sustainable use of biodiversity and ecosystems SDG 17: Partnership for the Goals – Number of science and/or technology cooperation agreements and programmes between countries (e.g., cooperation in the field of bioeconomy) – Total amount of approved funding for environmentally sound technologies (e.g., bio-based technologies) – Mechanisms in place to enhance the policy coherence of sustainable development		Competitiveness and creating jobs: – Bioeconomy in economy – Value and value-added – Exports – Comparative advantage – Employment in bioeconomy – Working conditions – Equality and inclusiveness – Physical infrastructures – Investments in rural and coastal areas – Rural income diversification – Income of primary producers – Knowledge on technologies – High-level education Research and innovation Market mechanisms Resource competition

Source: Author's processing

A comparative overview of criteria and indicators relevant for monitoring and evaluating the bioeconomy with the help of SDGs, the FAO approach, and the EC/JRS approach indicates the complexity of a holistic view of the bioeconomy. Given the range of generic indicators, the holistic evaluation of the bioeconomy is based on metadatabases that will provide interested analysts with the opportunity to choose and create their analytical frameworks for a

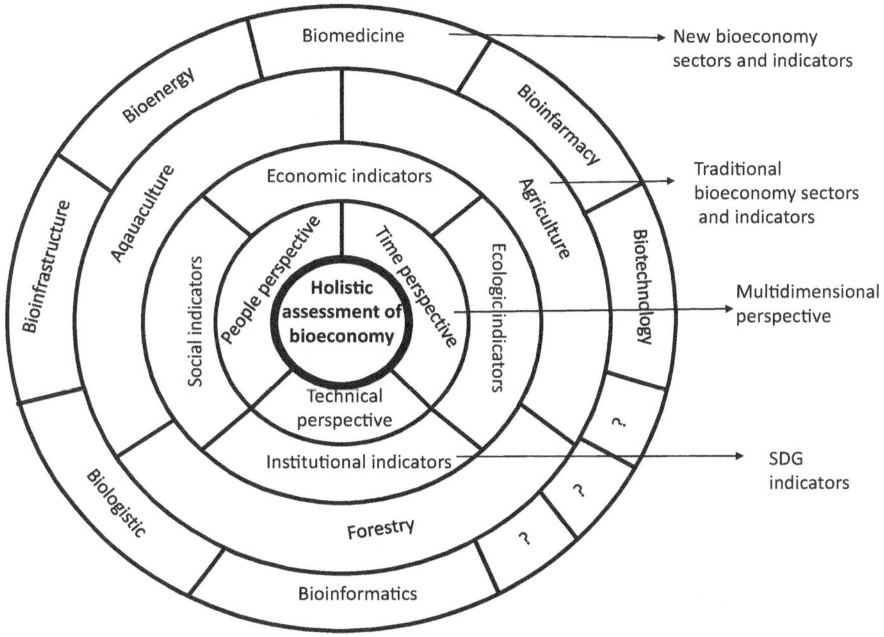

Figure 5.1 Complex interrelationships of indicators for monitoring the bioeconomy development

Source: Author

holistic view of the bioeconomy. The complex interrelationships of indicators that reflect the economy from different starting points are shown in Figure 5.1.

At the same time, interpreting trends still entails intensive work for the scientific community. Particular attention should be paid to identifying counter-effects, with an emphasis on the time dimension. Cross-cutting issues and counter-effects highlight the relevant interconnections between different goals, sectors, and criteria. Only by understanding the cross-cutting issues and counter-effects between them is it possible to provide a holistic assessment. This holistic assessment should avoid or minimise the counter-effects (ex-ante evaluation) or look for a remedy if the intervention has produced any damages (ongoing and ex-post evaluation).

Models for monitoring and evaluating the bioeconomy

Typical macroeconomic models adopted to measure the contribution of the bioeconomy to the development of an individual society include the value-added/GDP approach, Modular Applied GeNeral Equilibrium Tool (MAG-NET) and social accounting matrix analysis, computable general equilibrium model, partial equilibrium model, and other economic models and tools. However, such approaches do not systematically consider environmental and social aspects. Furthermore, the economic approach is characterised by certain limitations in considering contributions in the economic sphere resulting from the lack of a standard methodology that would allow an international comparison of the bioeconomy contribution to GDP. In addition, bioeconomy products and activities differ significantly in terms of a particular country's priorities and comparative advantages (Bracco *et al.* 2018).

The methodology for monitoring and evaluating the bioeconomy starts from a holistic aspect and includes qualitative and quantitative analysis of indicators regarding geographical features, sectoral division, time dimension, and people perspective.

Although an extensive database of statistical data is used to monitor the bioeconomy, an appropriate set of indicators, which may be presented in the form of an information pyramid, is needed to quantify key objectives. The pyramidal approach to the bioeconomy evaluation involves designing and monitoring indicators at three levels of aggregation: the base, middle, and top of the pyramid. Top indicators appear at the top of the pyramid, specific indicators in the middle, and the base includes fundamental indicators that are the basis for contemplating the bioeconomy. The main indicators are highly complex and aggregated to indicate crucial steps in the bioeconomy development and warn decision-makers of the need for a policy response. Therefore, they have a high degree of aggregation of a large amount of information and a very low level of detail. Specific indicators broaden and deepen the information provided by the main indicators. They are defined more narrowly but still retain the attributes of the indicators, which means that they adequately inform about certain characteristics of each main indicator. At the lowest level of the information pyramid, basic data provide information used to build specific and key indicators. These indicators are the simplest, the easiest to obtain when conducting an analysis, and are primarily used to holistically interpret indicators that are aggregated to a greater extent. At the same time, the information flow necessary for pyramidally creating and interpreting top-down and bottom-up indicators is essential in a holistic approach to the bioeconomy.

One of the recognised and used methods for monitoring and evaluating the development of bioeconomy providing holistic insight is the so-called life-cycle assessment (LCA) that aims to assess the potential environmental impacts of a product (Sevigné-Itoiz *et al.* 2021). This method involves collecting and analysing all possible impacts on the development of the bioeconomy, from the generation of ideas to the development of products or services and their consumption. The method seeks to cover all stages of the life cycle of ideas, technologies, products, or services in the field of bioeconomy and understand their sources, methods, rules, and consequences of design, use, and, ultimately, if appropriate, disposal. This method is fundamental in the field of bioeconomy because of the circularity that it implies. However, it does not clearly separate holistic observation and monitoring. The bioeconomy LCA is implemented in four phases:

1. Clearly defining the goal and scope of monitoring and evaluation with regard to the purpose of the results
2. Listing the desired system of indicators and ensuring the availability of data (e.g., input units can be energy, water, and raw materials; output units can be products, by-products of emissions [CO_2, CH_4, SO_2, No_x, CO] to water, soil, air, and solid waste)
3. Analysing the collected indicators through a selection of distinct categories of impacts, such as societal, economic, or environmental impacts and impacts on the adoption of public policies over time, in a certain territory, and among certain groups of stakeholders
4. Interpreting research findings, including the underlying conclusions, possible limitations in using the model and recommendations for the development of the bioeconomy, and ultimately inputs for relevant development policies

The LCA method emphasises the monitoring of the bioeconomy from the environmental, economic, and technical aspects. Still, it does not include a more substantial social dimension of bioeconomy development and a view through people and time perspectives. Over the past several years, the method has been supplemented with respect to social impacts (the so-called Social LCA or S-LCA). However, these social impacts have not been intensively applied in monitoring and evaluating the bioeconomy.

One of the methods used to understand better the impact of policies or technological developments on the bioeconomy is the MAGNET. This method is cross-sectoral in nature. It allows coherent simulation of possible alternative scenarios as it covers all economic sectors and regions worldwide and includes all relevant policies (with various levels of representation), thus enabling the more active involvement of technology partners in the analysis (Shutes *et al.* 2017). In this method, challenges relate to data availability, difficulty in establishing a connection with sector models, and complexity in data management (Shutes *et al.* 2017). It is also challenging to interpret the numerous and complex relationships between individual indicators and their mutual influences and interactions, which is generally one of the fundamental limitations in monitoring and evaluating the bioeconomy (Bracco *et al.* 2019).

Some limitations of the proposed methods for monitoring the bioeconomy are as follows:

- Sometimes it is difficult to establish a link between the chains of cause and effect, given the production activities and their potential social impacts, making it difficult to select appropriate indicators.

- The number and aggregation of indicators across all dimensions relevant for a holistic view of the bioeconomy is uneven. The development of quantitative indicators related to the social aspects of the bioeconomy still lags behind environmental or economic and technological indicators.
- The availability of individual indicators represents a particular problem as analyses mainly use publicly available data that have been found to be generally significant (e.g., gender equality, child labour, forced labour, access to clean energy), and not specific data that would address the bioeconomy at a more extensively aggregated level and the impact it has on the society in which it is developing.
- The harmonised data for the collection of specific indicators and ultimately robustness of results are lacking if the analyses include the analysis of scenarios and sensitivity analyses, affecting both the transparency of the findings and the inclusion of these findings in the recommendations for decision-makers.

Conclusion – future trends

The bioeconomy is a highly complex area in terms of monitoring and evaluation. The realisation and future growth of the bioeconomy are essential because they ultimately offer a pivotal contribution to achieving sustainable development. Even holistically speaking, the bioeconomy as a development paradigm has more significant global implications than the concept of sustainable development as it provides a resource for the survival of the planet as a whole. The very complexity of the concept, entailing new and old sectors, the geographic aspect, the time horizon aspect, and the involvement of interested stakeholders, indicates the need for careful consideration and design of indicators for monitoring and analytical methods for evaluating complex processes within the sector.

Every year, the international community's efforts in defining indicators are on the increase. Awareness of the need for metadatabases to map the systems contained in the bioeconomy and the selection of generic indicators that could be used at higher levels of analysis and for forecasting purposes are evolving daily. It is essential to have strong generic indicators that can monitor the development of the bioeconomy at a general level. Still, the complexity of eco-processes that take place in the processes of production and use of raw plant materials require a much broader holistic framework. The bioeconomy indicators should include additional criteria such as poverty reduction, gender equality, and health and safety in the bioeconomy. Therefore, in addition to the existing environmental and economic indicators, it is necessary to monitor and evaluate the social impacts of the bioeconomy development, where there is still much room for improvement.

Monitoring indicators and evaluating the bioeconomy is an ongoing process that should be based on publicly available and regularly updated data. Reliable data are needed to create reliable indicators in the form of a solid metadatabase and interdisciplinary knowledge. Indicators should cover all life-cycle aspects, including ecosystems and services provided. The development of new knowledge and investment in education are key steps for expanding on the system of bioeconomy indicators and their use in policy decision-making at the local, regional, national, and global levels. The cooperation between the FAO and the Commission's JRC is an excellent example of jointly developing a system of bioeconomy indicators.

Some of the guidelines for the future follow a holistic view in mapping the indicators throughout all key sectors that constitute the bioeconomy (traditional sectors such as agriculture, forestry, and aquaculture and new ones such as bioinformatics, biomedicine, bioenergy, etc.). They should enable economic, social, environmental, and institutional evaluation, include qualitative and quantitative insights across various levels of information generation, and incorporate a geographical, time, and people dimension.

A structured approach to selecting, collecting, and aggregating the bioeconomy indicators largely depends on the method used in the evaluation. For the past ten years, various analytical methods have been established for monitoring and evaluating bioeconomy development. Some of the methods represent an upgrade of existing tools that are used to monitor the development in individual economic sectors or national economies and that have been developed to support the development and follow-up of complex public policies. The more complex the models, the more complex the sensitivity analysis is based on different development factors. For example, including the technological aspect of development in the analysis may facilitate the identification of focal points of future optimisations and research efforts in the field of bioeconomy. In addition, evaluating advances in research and innovation in all sectors through the prism of the bioeconomy assesses the community's potential for creating new green jobs and green growth while reducing GHG emissions and making progress towards some of the sustainable development goals.

The methods to be used in future bioeconomy development assessments should certainly entail involving the entire value chain and all relevant stakeholders. For example, the logistics or municipal and transport infrastructure sectors may not yet be recognised as sectors supporting the development of the bioeconomy and, as such, are not included in models that monitor their development. Therefore, including the value chain certainly adds to the broader insight into traditional sectors and represents a way of encouraging the development of the bioeconomy. Furthermore, traditional monitoring and evaluation methods such as the pyramid model, life-cycle assessment and life-cycle sustainability assessment (LCSA) have practical applications in monitoring and evaluating the bioeconomy but face major methodological and practical issues. Therefore, developing a new model such as the holistic and integrated LCSA to be used in the integrated monitoring and evaluation of the bioeconomy, which would entail a single goal, all relevant functional units and variables, impact assessment, and associated interpretation, could describe social, economic, and environmental risks, opportunities, synergies, and costs (Zeug 2021).

The application of various analytical methods in monitoring and evaluating the bioeconomy development is ultimately intended to support policymaking, starting with defining policy programmes and helping define the objectives of complex, cross-sectoral, and interdisciplinary development policies. A structured approach to designing, selecting, collecting, and aggregating indicators, regardless of the method or analytical technique, allows for a comparative assessment of products, services, and processes; provides information on the use of resources; and gives an overview of past and future trends, influences, and effects of different social groups, which is the basis for a holistic view of future bioeconomy development.

Bibliography

Acevedo, T. A., Morales, M. R., and Lobato-Calleros, O., 2015. Additional indicators to promote social sustainability within government programs: Equity and efficiency. *Sustainability*, 7 (7), 9251–9267. Available from: https://doi.org/10.3390/su7079251 [Accessed 3 December 2021].

Beg, M., 2019. Održivi ekonomski razvoj – prepreke i rješenja, EFZG Occasional Publications (Department of Macroeconomics). In G. Družić and I. Družić, eds. *Zbornik radova znanstvenog skupa: Modeli razvoja hrvatskog gospodarstva*, 1st edn, vol. 1. Zagreb: Faculty of Economics and Business, University of Zagreb, Croatian Academy of Sciences and Arts, 371–394. Available from: http://web.efzg.hr/repec/Chapters/chapter18-16.pdf [Accessed 3 December 2021].

Bracco, S., Calicioglu, O., Gomez San Juan, M., and Flammini, A., 2018. Assessing the contribution of bioeconomy to the total economy: A review of national frameworks. *Sustainability*, 10 (6), 1698. Available from: http://dx.doi.org/10.3390/su10061698 [Accessed 3 December 2021].

Bracco, S., Tani, A., Çalıcıoğlu, Ö., Gomez San Juan, M., and Bogdanski, A., 2019. *Indicators to Monitor and Evaluate the Sustainability of Bioeconomy. Overview and a Proposed Way Forward.* Rome: FAO.

Calicioglu, O., and Bogdanski, A., 2021. Linking the bioeconomy to the 2030 sustainable development agenda: Can SDG indicators be used to monitor progress towards a sustainable bioeconomy? *New Biotechnology*, 61, 40–49. Available from: https://doi.org/10.1016/j.nbt.2020.10.010 [Accessed 3 December 2021].

Diaz-Chavez, R., 2014. Indicators for socio-economic sustainability assessment. In D. Rutz and R. Janssen, eds. *Socio-Economic Impacts of Bioenergy Production.* Cham: Springer International Publishing, pp. 17–37. Available from: https://doi.org/10.1007/978-3-319-03829-2_2 [Accessed 3 December 2021].

Diaz-Chavez, R., Stichnothe, H., and Johnson, K., 2016. Sustainability considerations for the future bioeconomy. In P. Lamers, E. Searcy, J. Richard Hess, and H. Stichnothe, eds. *Developing the Global Bioeconomy.* London: Academic Press, pp. 69–90.

Egenolf, V., and Bringezu, S., 2019. Conceptualisation of an indicator system for assessing the sustainability of the bioeconomy. *Sustainability*, 11 (2), 443. Available from: https://doi.org/10.3390/su11020443 [Accessed 3 December 2021].

European Bioeconomy Stakeholders, 2017. European bioeconomy stakeholders' manifesto. Available from: https://ec.europa.eu/info/files/european-bioeconomy-stakeholders-manifesto_en [Accessed 3 December 2021].

European Commission, 2015. *Sustainable agriculture, forestry and fisheries in the bioeconomy: A challenge for Europe: 4th SCAR foresight exercise*, In B. Kovacs, ed. Brussels: Publications Office. Available from: https://data.europa.eu/doi/10.2777/179843

European Commission, Joint Research Centre, Institute for Environment and Sustainability, 2010. *International Reference Life Cycle Data System (ILCD) Handbook – General Guide for Life Cycle Assessment – Detailed Guidance*. Luxembourg: Publications Office of the European Union.

FAO, 2016. *How Sustainability Is Addressed in Official Bioeconomy Strategies at International, National and Regional Levels: An Overview.* Rome: FAO.

Giuntoli, J., Robert, N., Ronzon, T., Sanchez Lopez, J., Follador, M., Girardi, I., Barredo Cano, J., Borzacchiello, M., Sala, S., M'Barek, R., La Notte, A., Becker, W., and Mubareka, S., 2020. *Building a Monitoring System for the EU Bioeconomy. Progress Report 2019: Description of Framework*. Luxembourg: Publications Office of the European Union.

Idrus, N., and Al-Amin, A. Q., 2016. *Developing an Index for Quantifying Economic Impacts: The Bioeconomy Contribution Index*. Kuala Lumpur: Malaysian Biotechnology Corporation (BiotechCorp).

Karvonen, J., Halder, P., Kangas, J., and Leskinen, P., 2017. Indicators and tools for assessing sustainability impacts of the forest bioeconomy. *Forest Ecosystems*, 4 (2). Available from: https://doi.org/10.1186/s40663-017-0089-8 [Accessed 3 December 2021].

Kilsedar, C., Girardi, I., Gerlach, H., and Mubareka, S., 2021. *Implementation of the EU Bioeconomy Monitoring System Dashboards*. Luxembourg: Publications Office of the European Union.

Melim-McLeod, C., Olsen, E., Follador, M., and Robert, N., 2022. *EU Bioeconomy Monitoring System Indicator Update 2021*. In S. Mubareka, ed. Luxembourg: Publications Office of the European Union.

Pfau, S. F., Hagens, J., Dankbaar, B., and Smits, A. J. M., 2014. Visions of sustainability in bioeconomy research. *Sustainability*, 6 (3), 1222–1249. Available from: https://doi.org/10.3390/su6031222 [Accessed 3 December 2021].

Ronzon, T., Piotrowski, S., Tamosiunas, S., Dammar, L., Carus, M., and M'Barek, R., 2020. Developments of economic growth and employment in bioeconomy sectors across the EU. *Sustainability*, 12 (11), 4507. Available from: https://doi.org/10.3390/su12114507 [Accessed 3 December 2021].

Sevigné-Itoiz, E., Mwabonje, O., Panoutsou, C., and Woods, J., 2021. Life cycle assessment (LCA): Informing the development of a sustainable circular bioeconomy? *Philosophical Transitions of the Royal Society A*, 379 (2206). Available from: https://doi.org/10.1098/rsta.2020.0352 [Accessed 3 December 2021].

Shutes, L., Philippidis, G., M'barek, R., Ronzon, T., Ferrari, E., and Van Meijl, H., 2017. SDGs in the global MAGNET model for policy coherence analysis, European Commission, 2017, Joint Research Centre no. 108799. Available from: https://publications.jrc.ec.europa.eu/repository/handle/JRC108799 [Accessed 3 December 2021].

Stiglitz, J. E., Sen, A., and Fitoussi, J.-P., 2009. The measurement of economic performance and social progress revisited, OFCE no. 33. Available from: www.ofce.sciences-po.fr/pdf/dtravail/WP2009-33.pdf [Accessed 3 December 2021].

United Nations, 2015. *Transforming our world: The 2030 agenda for sustainable development*. Available from: https://sustainabledevelopment.un.org/post2015/transformingourworld/publication [Accessed 3 December 2021].

Wesseler, J., and von Braun, J. 2017. Measuring the Bioeconomy: Economics and Policies. *Annual Review of Resource Economics*, 9:275–298. Available from: https://doi.org/10.1146/annurev-resource-100516-053701 [Accessed 3 December 2021].

Zeug, W., 2021. Towards a holistic and integrated life cycle sustainability assessment of the bioeconomy, holistic and integrated life cycle sustainability assessment of regional bioeconomy. Paper presented at the Third International Forest Policy Meeting (IFPM3), March 2021, University of Freiburg.

6 Conclusion

Sanja Tišma

As a modern-age economy, the bioeconomy represents all economic activities aimed at preserving natural resources and gaining reasonable control over the development and use of technologies to serve people's actual needs and improve living and working conditions. It includes a range of integrated sectors, services, and activities related to developing, producing, and using biological products and processes. Unlike global economic goals such as economic growth, security, stability, full employment, efficiency, and equity, the key objectives of the bioeconomy include low CO_2 emissions, sustainable production and consumption, and the protection of biodiversity and the environment. In this way, the bioeconomy can be seen as a result of the sustainable development concept, based on a responsible view on technologies, production processes, and the protection and preservation of the environment while reducing the depletion of natural resources. In a narrow sense, the bioeconomy is an innovative, low-emission economy that ensures the sustainability of agriculture and fisheries, food security, and the sustainable industrial use of renewable biological resources and biomass while protecting biodiversity and preserving the environment. A broader concept of the bioeconomy includes a series of new scientific disciplines and a number of breakthroughs contributing to the preservation of the planet, such as biomedicine, biopharmacy, bioenergy, bioeconomy, and the like.

This concept has been well-known since ancient times. Still, it was generally accepted as a global development paradigm in the early 2000s, when the bioeconomy was presented as part of the response to prevailing economic development challenges and the overriding productivity growth imperative in agricultural production. A strategic framework for the development of the bioeconomy at the global level has been developing for the past 90 years. National economies and local communities have been adopting a number of macro-regional bioeconomy strategies and development plans, as well as specific approaches to the bioeconomy. Although the background is similar, there are several differences in the understanding of the scope and in the areas of interest of the bioeconomy, which modifies the objectives of the activities and the measures proposed for the future. In addition, certain related development

DOI: 10.4324/9781003223733-7

concepts based on green and sustainable business, such as the green economy and the circular economy, have emerged over the past decade. The circular economy is a renewable system in which the input of resources and waste, emissions, and energy loss is minimised by slowing down, closing, and narrowing materials and energy loops. The green economy is a low-carbon, resource-efficient, and socially inclusive economy in which employment and income growth are driven by and achieved through public and private investment in economic activities, infrastructure, and assets that reduce carbon emissions and pollution, enhance energy and resource efficiency, and prevent the loss of biodiversity and ecosystem services.

The bioeconomy, the circular economy, and the green economy share the same goal of making development sustainable but differ in the level of comprehensiveness. The bioeconomy represents the narrowest concept based on bioproducts and bioenergy, and the green economy generally refers to renewable energy sources, organic production, recycling processes, and the social acceptability of economic activities. In contrast, the circular economy represents the broadest concept that includes all societal activities contributing to the preservation and protection of the environment.

The scope of the concept of bioeconomy is mainly related to the sectors it includes. All stakeholders involved agree that the traditional sectors of the bioeconomy are agriculture, forestry, and fisheries (i.e., aquaculture). Agriculture is a sector that is essential for life and food security, but it also forms the basis for the development of local entrepreneurship, job creation, employment, and social development. Agriculture's impact on the development of the bioeconomy is based on the use of renewable resources to produce food, materials, and energy, making resource efficiency and the transition to a low-carbon economy possible. Forestry participates in the scope of the bioeconomy through forest products, wood pulp, raw materials, and wood-based products. The so-called blue bioeconomy, which uses microorganisms, algae, and invertebrates as water biomass, is linked to the fisheries and aquaculture sector.

New bioeconomy sectors represent developmental breakthroughs for the future. They include the development of new technologies based on bio raw materials and bioprocesses such as biomedicine, biopharmacy, biotechnology, bioinformatics, bioenergy, and the like. Biomedicine, for example, includes the development of precision medicine and regenerative medicine, which represent entirely new ways of treating disease. Biopharmaceuticals play an essential role in promoting global health through herbal drug production. They have significant development potential and contribute to improving the quality of life of all living beings globally. Biopharmaceutical technology is recognised as one of the key strategic technologies in the 21st century. In addition, bioenergy could significantly impact the fight against climate change, while bioinformatics is becoming increasingly important in the world of metadatabases.

Awareness of the future importance of the bioeconomy has influenced global initiatives to monitor the development of this area and its impact on global development. Therefore, the wider international community has intensified its efforts in developing indicators to monitor the bioeconomy and its implications for economic growth and development, primarily on the labour market and the creation of new green jobs, over the past few years. Monitoring indicators and assessing the bioeconomy is an ongoing process that should be based on publicly available and regularly updated data. Reliable data are needed to create reliable indicators in the form of a solid knowledge base through all four pillars of sustainability, namely the environmental, economic, social, and political/managerial, thus establishing a holistic aspect in evaluating the bioeconomy. Regular and comprehensive monitoring of the bioeconomy is needed so that policymakers and stakeholders can check whether policy implementation is "on track" and generate information that can be used to assess the achievement of their goals. A holistic approach is necessary to get a complete picture of bioeconomy trends and their impact on society.

Globally, the world is facing an increasing demand for limited natural resources and their impact on living sustainability on planet Earth. Therefore, the critical goals for the future development of the bioeconomy are related to harmonisation of sectoral public policies, regulation of the legislative framework in the field of bioeconomy, development of the education system, the fostering of scientific research, international cooperation, and clear and intensive communication with the general public. This is especially important in light of climate change and the Covid-19 pandemic. As a paradigm, the bioeconomy is more relevant now than ever before. The development of the bioeconomy entails investments in new environmentally friendly technologies, clean production, new green products, and socially responsible behaviour of all stakeholders. Finally, the bioeconomy can also be a driver of fundamental and systemic change in establishing new ways of producing and consuming biological resources while respecting the ecological boundaries of our planet. At the same time, a holistic approach to considering and understanding the challenges and potentials of the bioeconomy represents an innovation in everyday practice.

The message conveyed throughout the book is that the bioeconomy is an extremely valuable tool in combating various current global issues, including population growth, poverty growth, environmental challenges and problems, climate change, and many more. As a concept based on using natural resources in production, the bioeconomy primarily tackles environmental issues and indirectly deals with other social and economic problems.

The importance of the bioeconomy in the fight against current global issues has been recognised worldwide, which is why so many world powers have implemented strategies to develop the bioeconomy. Smaller countries have followed their example; today, many nations worldwide have their bioeconomy strategies. However, bioeconomy strategies worldwide differ according to a country's stage of development, its definition of the bioeconomy and sectors it encompasses, and the very desires and needs in shifting the economy from fossil to renewable sources.

The bioeconomy shows enormous potential in the energy, agricultural, food, and feed sectors. Furthermore, new products such as chemicals based on renewable energy sources show great promise concerning the transition from a fossil economy to a bioeconomy. In addition, new technologies such as genome editing in plant cultivation, metabolic engineering, and further digitisation of the primary sector are paving the way for furthering the development of the bioeconomy. The trust in bioeconomy-based innovation has been on the increase in view of the need to develop new bio-based products and optimise agriculture to feed a growing world population. The bioeconomy provides tools to address the challenge of increasing yields and optimising land use with the help of technological progress. Furthermore, the bioeconomy offers tools for achieving Sustainable Development Goals.

The future of the bioeconomy also depends on the application of biotechnology, which has the potential to improve and manage food, feed, and fibre crops and is driven by growing demand and increased agronomic stress due to climate change. Furthermore, the expected long-term increase in the price of fossil fuels due to the decline in the supply of low-cost oil, as well as the increase in demand for energy, and restrictions on greenhouse gas (GHG) production, constitute the basis for the growth of the biomass market, including non-food crops such as grass and wood, as raw materials for biofuels, chemicals, and plastics. The bioeconomic potential of biotechnology lies in the use of plants to produce valuable chemicals such as biopharmaceuticals and the production of nutraceuticals from plant and animal sources. The projected increase in demand for biomass for food and industrial purposes will lead to the need for a sustainable increase in agricultural, forestry, fisheries, and aquaculture capacities. On the other hand, the development of production systems with reduced GHGs emissions will lead to the adaptation and mitigation of the harmful effects of climate change, such as droughts and floods, but also contribute to the transition to a low-carbon economy with efficient resources.

To make use of the potential of the bioeconomy, it is necessary to manage natural resources wisely and sustainably and tackle the challenges of developing the bioeconomy, but also take regional and local aspects into account when preparing the bioeconomy strategies. Existing policies need to be carefully complemented, bearing in mind that the needs of less developed and developing countries are technologically different from the needs of industrialised, highly developed countries. Future efforts should certainly include educational, financial, and policy measures to promote a thriving bioeconomy market. Knowledge and funding are vital to improving the bioeconomy in the future. National strategies should prioritise sufficient funding and access to capital, as well as training programmes for professionals and the general population. It will also be necessary to pool knowledge between the industrialised and developing countries to ensure positive developments in the bioeconomy.

Index

Note: Page numbers in *italics* indicate a figure and page numbers in **bold** indicate a table on the corresponding page.